T0324775

VOLUME TWO HUNDRED AND TWO

Advances in
IMAGING AND
ELECTRON PHYSICS

EDITOR-IN-CHIEF

Peter W. Hawkes
CEMES-CNRS
Toulouse, France

VOLUME TWO HUNDRED AND TWO

ADVANCES IN
IMAGING AND
ELECTRON PHYSICS

Edited by

PETER W. HAWKES
CEMES-CNRS
Toulouse, France

ACADEMIC PRESS
An imprint of Elsevier

Cover photo credit:
The Cover picture is taken from Eq. (53) of the chapter by Clifford M. Krowne in this volume.

Academic Press is an imprint of Elsevier
125 London Wall, London EC2Y 5AS, United Kingdom
525 B Street, Suite 1800, San Diego, CA 92101-4495, United States
50 Hampshire Street, 5th Floor, Cambridge, MA 02139, United States
The Boulevard, Langford Lane, Kidlington, Oxford OX5 1GB, United Kingdom

Notices

Knowledge and best practice in this field are constantly changing. As new research and experience broaden our understanding, changes in research methods, professional practices, or medical treatment may become necessary.

Practitioners and researchers must always rely on their own experience and knowledge in evaluating and using any information, methods, compounds, or experiments described herein. In using such information or methods they should be mindful of their own safety and the safety of others, including parties for whom they have a professional responsibility.

To the fullest extent of the law, neither the Publisher nor the authors, contributors, or editors, assume any liability for any injury and/or damage to persons or property as a matter of products liability, negligence or otherwise, or from any use or operation of any methods, products, instructions, or ideas contained in the material herein.

ISBN: 978-0-12-812088-0
ISSN: 1076-5670

For information on all Academic Press publications
visit our website at https://www.elsevier.com/books-and-journals

**Working together
to grow libraries in
developing countries**

www.elsevier.com • www.bookaid.org

Publisher: Zoe Kruze
Acquisition Editor: Poppy Garraway
Editorial Project Manager: Shellie Bryant
Production Project Manager: Magesh Kumar Mahalingam
Designer: Mark Rogers

Typeset by VTeX

CONTENTS

3. The Struggle to Overcome Spherical Aberration in Electron Optics   75

Albert Septier

CONTRIBUTORS

Jesús Angulo
CMM – Centre de Morphologie Mathématique, MINES ParisTech, PSL Research University, Paris, France

Clifford M. Krowne
Naval Research Laboratory, Washington, DC, United States

Albert Septier (1924–2012)
Institute of Electronics, Faculty of Sciences of Orsay, University of Paris, Orsay, France

Santiago Velasco-Forero
CMM – Centre de Morphologie Mathématique, MINES ParisTech, PSL Research University, Paris, France

PREFACE

Mathematical morphology regularly makes its appearance in these Advances and several contributions on the subject are planned for the next few volumes. The first chapter here, by J. Angulo and S. Velasco-Forero, explains an important recent development, non-negative sparse morphology. This is a consequence of work on sparse modeling of images, which is vital for such large datasets. How can such sparse representations be used to calculate approximations to the morphological operators? It is this question that is addressed by the authors, with many examples.

The second chapter brings us to electron physics and in particular to the relation between the critical temperature of a superconductor and structural disorder. This work by C.M. Krowne is especially relevant to the design of components for nanoscience. A further contribution on a related topic will appear in the next volume.

We conclude with a further chapter in our program of reprinting major articles from Advances in Optical and Electron Microscopy. Here, we reproduce a heavily cited essay by the late Albert Septier on 'The struggle to overcome spherical aberration in electron optics'. This was for many years the authoritative account of work on the various ways of combating this resolution-limiting aberration of electron lenses and it has the additional merit of retrieving from oblivion several less well-known endeavors.

I am most grateful to the authors for their efforts to make unfamiliar material accessible.

Peter W. Hawkes

FUTURE CONTRIBUTIONS

S. Ando
Gradient operators and edge and corner detection

J. Angulo, S. Velaso-Forero
Convolution in (max, min)-algebra and its role in mathematical morphology

D. Batchelor
Soft x-ray microscopy

E. Bayro Corrochano
Quaternion wavelet transforms

C. Beeli
Structure and microscopy of quasicrystals

C. Bobisch, R. Möller
Ballistic electron microscopy

F. Bociort
Saddle-point methods in lens design

K. Bredies
Diffusion tensor imaging

A. Broers
A retrospective

A. Cornejo Rodriguez, F. Granados Agustin
Ronchigram quantification

J. Elorza
Fuzzy operators

R.G. Forbes
Liquid metal ion sources

P.L. Gai, E.D. Boyes
Aberration-corrected environmental microscopy

S. Golodetz
Watersheds and waterfalls

R. Herring, B. McMorran
Electron vortex beams

F. Houdellier, A. Arbouet
Ultrafast electron microscopy

M.S. Isaacson
Early STEM development

K. Ishizuka
Contrast transfer and crystal images

K. Jensen, D. Shiffler, J. Luginsland
Physics of field emission cold cathodes

U. Kaiser
The sub-Ångström low-voltage electron microscope project (SALVE)

K. Kimoto
Monochromators for the electron microscope

O.L. Krivanek
Aberration-corrected STEM

M. Kroupa
The Timepix detector and its applications

C. Krowne
Critical magnetic field and its slope, specific heat and gap for superconductivity as modified by nanoscopic disorder

B. Lencová
Modern developments in electron optical calculations

H. Lichte
Developments in electron holography

M. Matsuya
Calculation of aberration coefficients using Lie algebra

J.A. Monsoriu
Fractal zone plates

L. Muray
Miniature electron optics and applications

M.A. O'Keefe
Electron image simulation

V. Ortalan
Ultrafast electron microscopy

D. Paganin, T. Gureyev, K. Pavlov
Intensity-linear methods in inverse imaging

N. Papamarkos, A. Kesidis
The inverse Hough transform

H. Qin
Swarm optimization and lens design

Q. Ramasse, R. Brydson
The SuperSTEM laboratory

B. Rieger, A.J. Koster
Image formation in cryo-electron microscopy

P. Rocca, M. Donelli
Imaging of dielectric objects

J. Rodenburg
Lensless imaging

J. Rouse, H.-n. Liu, E. Munro
The role of differential algebra in electron optics

J. Sánchez
Fisher vector encoding for the classification of natural images

P. Santi
Light sheet fluorescence microscopy

R. Shimizu, T. Ikuta, Y. Takai
Defocus image modulation processing in real time

T. Soma
Focus–deflection systems and their applications

J. Valdés
Recent developments concerning the Système International (SI)

J. van de Gronde, J.B.T.M. Roerdink
Modern non-scalar morphology

CHAPTER ONE

Non-Negative Sparse Mathematical Morphology

Jesús Angulo[1], Santiago Velasco-Forero
CMM – Centre de Morphologie Mathématique,
MINES ParisTech, PSL Research University, Paris, France
[1]Corresponding author: e-mail address: jesus.angulo@mines–paristech.fr

Contents

1. INTRODUCTION

Modern image processing techniques should deal with images datasets of huge volume, great variety (e.g., optical, radar and hyperspectral sensors, sequence of images often multi–scale, etc.) and different veracity (i.e., incomplete images, or different uncertainty in the acquisition procedure). This paper addresses the problem of simultaneous treatment of a set of images via mathematical morphology. Our approach is motivated by sound developments in sparse signal representation (or *coding*), which

1

suggest that the linear relationships among high–resolution signals can be accurately recovered from their low–dimensional projections (also called *dictionary*) (Donoho, 2006). Sparse coding and dictionary learning, where data is assumed to be well represented as a linear combination of a few elements from a dictionary, is an active research topic which leads to state–of–the–art results in image processing applications, such as image denoising, inpainting, demosaicking or compression (Elad & Aharon, 2006; Mairal, Elad, & Sapiro, 2008; Yu, Sapiro, & Mallat, 2011). For a detailed self–contained overview on sparse modeling for image and vision processing the reader is referred to Mairal, Bach, and Ponce (2012). Inspired by this paradigm of parsimony representation of images, the aim of this work is to explore how image sparse representations can be useful to efficiently calculate approximations to morphological operators.

Mathematical morphology (Serra, 1982, 1988; Heijmans, 1994) is a nonlinear image processing methodology based on the application of lattice theory to spatial structures. Morphological filters and transformations are useful for various image processing tasks (Soille, 1999), such as denoising, contrast enhancement, multi–scale decomposition, feature extraction and object segmentation. In addition, morphological operators are defined using very intuitive geometrical notions which allows us the perceptual development and interpretation of complex algorithms by combination of various operators.

Notation. Let E be a domain of points, which is considered here as a finite digital space of the pixels of the image, i.e., $E \subset \mathbf{Z}^2$ such that $N = |E|$ is the number of pixels. Image intensities are numerical values, which ranges in a closed subset $\mathcal{T}$ of $\overline{\mathbf{R}} = \mathbf{R} \cup \{-\infty, +\infty\}$; for example, for an image of discrete L values, it can be assumed $\mathcal{T} = \{t_1, t_2, \cdots, t_L\}$. A binary image X is modeled as a subset of E, i.e., $X \in \mathcal{P}(E)$; a gray-level image $f(p_i)$, where $p_i \in E$ are the pixel coordinates, is a numerical function $E \to \mathcal{T}$, i.e., $f \in \mathcal{F}(E, \mathcal{T})$. In this paper, we are interested in operators ψ as a map transforming an image into an image. There are thus operators on binary images, i.e., maps $\mathcal{P}(E) \to \mathcal{P}(E)$; or on gray-level images, i.e., maps $\mathcal{F}(E, \mathcal{T}) \to \mathcal{F}(E, \mathcal{T})$.

Motivation and outline of the approach. In order to illustrate our motivation, let us consider the diagrams depicted in Fig. 1. In many practical situations, a collection $\mathcal{F} = \{f_1(p_i), \cdots, f_M(p_i)\}$ of M binary or gray-level images (each image having N pixels) should be analyzed by applying the same

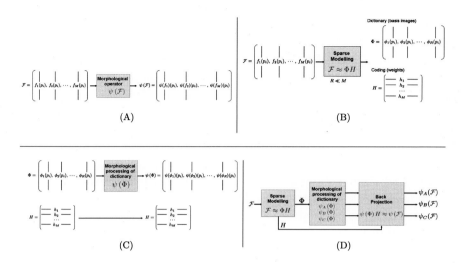

Figure 1 Sparse mathematical morphology: motivation and outline.

morphological operator ψ (or a series of operators) to each image of $\mathcal{F}$, i.e., $\psi(\mathcal{F}) = \{\psi(f_j)\}_{1 \leq j \leq M}$, Fig. 1A. If one considers that the content of the various images is relatively similar, we can expect that the initial collection can be efficiently projected into an R dimensionality reduced image space, using typically sparse modeling techniques, i.e., $\mathcal{F} \approx \Phi H$, in order to learn the dictionary composed of atom images $\Phi = \{\phi_1(p_i), \cdots, \phi_R(p_i)\}$ and the corresponding coding or weights H, Fig. 1B. In this context, we would like to apply the morphological operator ψ (or an equivalent operator) to the reduced set of images of the projective space Φ, i.e., $\psi(\Phi) = \{\psi(\phi_k)\}_{1 \leq k \leq R}$, in such a way that the original processed images are approximately obtained by projecting back to the initial space, i.e., $\psi(\mathcal{F}) \approx \psi(\Phi) H$, Fig. 1C. Note that if the number of operators to be applied increases, e.g., Φ_A, Φ_B, Φ_C, etc., then such paradigm becomes much more efficient since $R \ll M$, Fig. 1D. At this point, we should remark that our approach is different from the standard use of sparse modeling in image processing (Mairal et al., 2012), since in the classical approaches the dictionary is not processed. In fact, sparse representation itself leads to a solution for the tasks of denoising, regularization, coding, etc.

Typical examples of image families which can be fit in our framework are: (i) a collection of binary shapes, (ii) a database of registered gray scale images, for instance, faces, cells (from biomicroscopy applications) or galaxies (in astronomy), (iii) a set of patches of a large image, (iv) a time series of registered images, (v) the spectral bands of a multi/hyper–spectral image, etc. Some of these examples are used to illustrate the relevance of our approach. The rationale behind this methodology is that the *intrinsic dimension*[1] of the image collection is lower than M. Usually the subspace representation involves deriving a set of basis components using linear techniques like Principal Component Analysis (PCA), Independent Component Analysis (ICA), or sparse dictionary learning techniques like K–SVD (Elad & Aharon, 2006). The projection coefficients for the linear combinations in the above methods can be either positive or negative, and such linear combinations generally involve complex cancellations between positive and negative numbers. Therefore, these representations lack the intuitive meaning of "adding parts to form a whole". This property is particularly problematic in the case of mathematical morphology since the basic binary operator, the dilation of a set, is defined as the operator which commutes with the union of parts of the set. This principle is widely discuss in the paper. In practice, that means that the paradigm discussed above in Fig. 1 cannot be used with sign unconstrained sparse modeling methods. Non–negative matrix factorization (NMF) (Lee & Seung, 1999) imposes the non–negativity constraints in learning basis images: the pixels values of resulting images as well as the coefficients for the reconstruction are all non–negative. This ensures that NMF is a procedure for learning a parts-based representation (Lee & Seung, 1999). In addition, by incorporating sparse constraint to NMF, it will be constructed a succinct representation of the image data as a combination of a few typical patterns (few atoms of the dictionary) learned from the data itself. Hence, our approach of sparse mathematical morphology is founded on sparse NMF modeling.

NMF state–of–the–art. The state–of–the–art on NMF is nowadays vast. Besides the pioneering work (Lee & Seung, 1999), several NMF variants have been developed by introducing additional constraints and properties to the original NMF. In particular, being relevant for our purpose, we can mention, on the one hand, the local NMF (Li, Hou, Zhang, & Cheng, 2001), which learns spatially localized, parts-based subspaces for images (i.e., visual patterns); and on the other hand, the sparse NMF (Hoyer, 2004), which incorporates explicit sparseness constraints. These powerful and already classical algorithms are reviewed in next section. We note also

---

[1] Roughly, the intrinsic dimension of an image f is defined as the number of "independent" parameters needed to represent f.

that in hyperspectral imaging, there is a common task called *linear unmixing* (or factorization into physical space) which consists in detecting the spectra of the pure materials and estimating their relative abundances. Linear unmixing can be seen as an equivalent problem to NMF (Esser, Möller, Osher, Sapiro, & Xin, 2012).

Formulation of PCA-like optimization with nonnegative and sparse constraints have also considered by several works, see for instance Zass and Shashua (2006). Motivated by a combination of NMF and kernel theory (embedding into polynomial feature space), various nonlinear variants of NMF have been proposed (Zafeiriou & Petrou, 2010). Similarly, algorithms have been also introduced for nonnegative ICA (Plumbley, 2003). NMF has been extended to factorization of non-vector spaces, typically for data sampled from a low-dimensional manifold embedded in a high-dimensional space, using affinity graph matrix non-negative decomposition (Cai, He, Han, & Huang, 2011) and manifold regularized NMF (Guan, Tao, Luo, & Yuan, 2011). NMF has been also extended to tensor factorizations (Cichocki, Zdunek, Phan, & Amari, 2009), and has considered for applications to exploratory multi-way data analysis and blind source separation. In general, NMF is a NP-hard problem (Vavasis, 2009). However, starting from the notion of separable NMF, introduced by Donoho and Stodden (2004), which in geometric terms involves that the conical hull of a small subsets of the columns of the data matrix contain the rest of the columns, an emerging line of work in NMF, based on computational geometry approaches, is leading to efficient solutions. It can be then solved by linear programming (Arora, Ge, Kannan, & Moitra, 2012), conical hull algorithms (Kumar, Sindhwani, & Kambadur, 2013) and random projections (Damle & Sun, 2014). When the data to be factorized is binary or Boolean (i.e., only "0" or "1" values), the NMF problem is called discrete basis problem (Miettinen, Mielikainen, Gionis, Das, & Mannila, 2008). This discrete basis problem which is equivalent to the binary k-median problem is NP-hard and some greedy algorithms (Miettinen, 2010) have been proposed to deal with. Other binary matrix factorization approaches are straightforward related to classical NMF (Zhang, Ding, Li, & Zhang, 2007) either by a thresholding algorithm (smooth approximation to Heaviside function) or by binary-constraint penalty function. A binary version of the sparse NMF can be also mentioned (Yuan, Li, Pang, Lu, & Tao, 2009), based on windowed image parts with Haar-like box functions. Extension of NMF to binary data factorization has been also considered by means of a Bernoulli distribution of binary observations (Schachtner, Pöppel, & Lang, 2010), or more recently using a logistic PCA approach (Tomé, Schachtner, Vigneron, Puntonet, & Lang, 2015).

Paper organization. The methodology described here was initiated in Angulo and Velasco-Forero (2011) for binary erosion/dilation. In the present work, we develop also sparse approximations to binary opening/closing as well as the full counterpart of sparse approximation to morphological operators for gray-scale images. The applications considered in the last part of the paper are also new with respect to our previous work (Angulo & Velasco-Forero, 2011).

The rest of paper is structured as follows. Section 2 reviews the formal definition of NMF and various algorithms proposed in the state-of-the-art, including the sparse variant introduced in Hoyer (2004), which is the one used in the our study. The use of NMF representations for implementing sparse pseudo-morphological binary operators is introduced and theoretically justified in Section 3. The section discusses also why sparse NMF is appropriate in morphological image processing. Section 4 generalizes the framework of morphological sparse processing to gray-scale images using two alternatives. The first one is based on level set decomposition of numerical functions. The second approach, more useful in practice, uses a straightforward representation of sparse NMF basis from the gray-scale images. Applications are studied in Section 5. The first case-study deals with the sparse processing and modeling of a multivariate Boolean texture. The second application focuses on morphological sparse processing of hyperspectral images. Conclusions and perspectives close the paper in Section 6.

2. NMF AND SPARSE VARIANTS

2.1 Definition of NMF on Vector Space

Let us assume that our data consists of M vectors of N non-negative scalar variables. Denoting the column vector $\mathbf{v}_j$, $j = 1, \cdots, M$, the matrix of data is obtained as $\mathbf{V} = (\mathbf{v}_1, \cdots, \mathbf{v}_M)$ (each $\mathbf{v}_j$ is the j-th column of $\mathbf{V}$), with $|\mathbf{v}_j| = N$. If we analyze M images of N pixels, these images can be stored in linearized form, so that each image will be a column vector of the matrix.

Given the non-negative matrix $\mathbf{V} \in \mathbb{R}^{N \times M}$, $\mathbf{V}_{i,j} \geq 0$, NMF is defined as a linear non-negative approximate data decomposition into the two matrices $\mathbf{W} \in \mathbb{R}^{N \times R}$ and $\mathbf{H} \in \mathbb{R}^{R \times M}$ such that

$$\mathbf{V} \approx \mathbf{WH}, \quad \text{s.t. } \mathbf{W}_{i,k}, \mathbf{H}_{k,j} \geq 0, \tag{1}$$

where usually $R \ll M$, which therefore involves a dimensionality reduction. Each of the R columns of $\mathbf{W}$ contains a basis vector $\mathbf{w}_k$ and each row of $\mathbf{H}$ contains the coefficient vector (weights) $\mathbf{h}_j$ corresponding to vector $\mathbf{v}_j$: $\mathbf{v}_j = \sum_{k=1}^{R} \mathbf{w}_k \mathbf{H}_{k,j} = \mathbf{W} \mathbf{h}_j$. Using the modern terminology from sparse coding theory, the matrix $\mathbf{W}$ contains the dictionary and $\mathbf{H}$ the encoding.

A theoretical study of the properties of NMF representation has been achieved in Donoho and Stodden (2004) using geometric notions. Hence, NMF is interpreted as the problem of finding a simplicial cone which contains the data points in the positive orthant, or in other words, NMF is a conical coordinate transformation. It is interesting to note that NMF is somehow equivalent to Kernel k-means clustering and Laplacian-based spectral clustering (Ding, He, & Simon, 2005).

2.2 Basic NMF Algorithms

The factorization $\mathbf{V} \approx \mathbf{W} \mathbf{H}$ is not necessarily unique, and the optimal choice of matrices $\mathbf{W}$ and $\mathbf{H}$ depends on the cost function that minimizes the reconstruction error. The most widely used is the Euclidean distance, i.e., minimize

$$\|\mathbf{V} - \mathbf{W} \mathbf{H}\|_2^2 = \sum_{i,j} \left(\mathbf{V}_{i,j} - (\mathbf{W} \mathbf{H})_{i,j} \right)^2,$$

with respect to $\mathbf{W}$ and $\mathbf{H}$, and subject to the constraints $\mathbf{W}, \mathbf{H} > 0$. Although the minimization problem is convex in $\mathbf{W}$ and $\mathbf{H}$ separately, it is not convex in both simultaneously. In Lee and Seung (2001) a multiplicative good performance algorithm to solve (1) is proposed. They proved that the cost function is nonincreasing at the iteration and the algorithm converges at least to a local optimal solution. More precisely, the update rules for both matrices are:

$$\mathbf{H}_{k,j} \leftarrow \mathbf{H}_{k,j} \frac{(\mathbf{W}^T \mathbf{V})_{k,j}}{(\mathbf{W}^T \mathbf{W} \mathbf{H})_{k,j}}; \quad \mathbf{W}_{i,k} \leftarrow \mathbf{W}_{i,k} \frac{(\mathbf{V} \mathbf{H}^T)_{i,k}}{(\mathbf{W} \mathbf{H} \mathbf{H}^T)_{k,j}}.$$

Another useful cost function, also considered in Lee and Seung (2001), is the Kullback–Leibler (KL) divergence, which leads also quite simple multiplicative update rules.

In Li et al. (2001), a variant of KL divergence NMF was proposed, which is named Local NMF (LNMF), aiming at learning spatially localized components (by minimizing the number of basis R to represent $\mathbf{V}$ and by maximizing the energy of each retained components) as well as imposing that different bases should be as orthogonal as possible (in order to minimize redundancy between the different bases). The multiplicative update rules for LNMF are given by

$$\mathbf{H}_{k,j} \leftarrow \sqrt{\mathbf{H}_{k,j} \sum_i \mathbf{V}_{i,j} \frac{\mathbf{W}_{i,k}}{(\mathbf{WH})_{i,k}}}; \quad \mathbf{W}_{i,k} \leftarrow \mathbf{W}_{i,k} \frac{\sum_j \mathbf{V}_{i,j} \frac{\mathbf{H}_{k,j}}{(\mathbf{WH})_{i,j}}}{\sum_j \mathbf{H}_{k,j}};$$

$$\mathbf{W}_{i,k} \leftarrow \frac{\mathbf{W}_{i,k}}{\sum_i \mathbf{W}_{i,k}}.$$

As we have discussed in the introduction, there are many other alternative NMF algorithms, but we focus on the following technique which provide a sparse variant.

2.3 NMF with Sparseness Constraints

A very powerful framework to add an explicit degree of sparseness in the basis vectors $\mathbf{W}$ and/or the coefficients $\mathbf{H}$ was introduced in Hoyer (2004). First of all, the sparseness measure σ of a vector $\mathbf{v} \in \mathbb{R}^{N \times 1}$ used in Hoyer (2004) is based on the relationship between the L_1 norm and the L_2 norm:

$$\sigma(\mathbf{v}) = \frac{\sqrt{N} - \|\mathbf{v}\|_1/\|\mathbf{v}\|_2}{\sqrt{N} - 1}.$$

This function is maximal at one if and only if $\mathbf{v}$ contains only a single non-zero component, and takes a value of zeros if and only if all components are equal (up to signs). Then, matrix $\mathbf{W}$ and $\mathbf{H}$ are solved by the problem (1) under additional constraints $\sigma(\mathbf{w}_k) = S_w$ and $\sigma(\mathbf{h}_j) = S_h$, where S_w and S_h are respectively the desired sparseness of $\mathbf{W}$ and $\mathbf{H}$. The algorithm introduced in Hoyer (2004) is a projected gradient descent algorithm (additive update rule), which takes a step in the direction of the negative gradient

$$\mathbf{W}_{i,k} \leftarrow \mathbf{W}_{i,k} - \mu_{\mathbf{W}} \left(\mathbf{W}_{i,k}\mathbf{H}_{k,j} - \mathbf{V}_{i,j}\right) \mathbf{H}_{j,k}^T;$$
$$\mathbf{H}_{k,j} \leftarrow \mathbf{H}_{k,j} - \mu_{\mathbf{H}} \mathbf{W}_{k,i}^T \left(\mathbf{W}_{i,k}\mathbf{H}_{k,j} - \mathbf{V}_{i,j}\right);$$

and subsequently projects each column of $\mathbf{W}$ and each row of $\mathbf{H}$ onto the constraint space. The most sophisticated step is therefore how to find, for a given vector $\mathbf{v}$, the closest non-negative vector $\mathbf{u}$ with a given L_1 norm and a given L_2 norm. The algorithm works as follows (Hoyer, 2004). The vector is projected onto the hyperplane of L_1. Next, within this space, one projects to the closest point on the joint constraint hypersphere (intersection of the sum and the L_2 constraints), by moving radially outward for the center of the sphere (the center is the point where all the components have equal values). If the components of this projection point are not completely non-negative, the negative values must be fixed at zero and a new point found in similar way under the same constraints. Sparseness is controlled explicitly with a pair of parameters that is easily interpreted; in addition, the number of required iterations grows very slowly with the dimensionality of the problem. In fact, for all the empirical tests considered in this paper, we have used the MATLAB code for performing NMF and its various extensions (LNMF, sparse NMF) provided by Hoyer (2004).

Besides the sparseness parameters (S_w, S_h), a crucial parameter to be chosen in any NMF algorithm is the value of R, that is, the number of basis of projective reduced space. Any dimensionality reduction technique, such as PCA, requires also to fix the number of components. In PCA, the components are ranked according to the second-order statistical importance of the components and each one has associated a value of the represented variance; whereas in NMF the selection R can be evaluated only *a posteriori*, by evaluating the error of reconstruction. This issue is not considered in the paper and we empirically fix R.

2.4 A Few Properties of NMF

We conclude this review on NMF algorithms by discussing some well-known properties which are useful for the sequel.

- **Boundedness property (for any NMF variant)** (Zhang et al., 2007): We say that $\mathbf{V}$ is bounded if $0 \leq \mathbf{V}_{i,j} \leq 1$. If $\mathbf{V}$ is bounded, then the factor matrices $\mathbf{W}$ and $\mathbf{H}$ are also bounded, i.e., $0 \leq \mathbf{W}_{i,k} \leq 1$ and $0 \leq \mathbf{H}_{k,j} \leq 1$.

- **Indeterminacies (for any NMF variant)** (Theis, Stadlthanner, & Tanaka, 2005; Huang, Sidiropoulos, & Swami, 2014): Positivity and sparseness are invariant under permutation and scaling of columns of $\mathbf{W}$ (and correspondingly of the rows of $\mathbf{H}$), i.e., $\mathbf{V} = \mathbf{WH} = (\mathbf{WP}^{-1}\mathbf{L}^{-1})(\mathbf{LPH})$, where $\mathbf{P}$ is a permutation matrix and $\mathbf{L}$ a positive scaling matrix.

- **Uniqueness (only for Sparse-NMF)** (Theis et al., 2005; Huang et al., 2014): Under sparsity constraints, projection step proposed in Hoyer (2004) has a unique solution, which is found. Hence, under non-degenerate data, Sparse-NMF a.s. produces a unique factorization.

This is another fundamental reason why we think that Sparse–NMF is an excellent choice for our learning our non–negative dictionaries.

3. SPARSE APPROXIMATION TO BINARY MORPHOLOGICAL OPERATORS

Let $\mathcal{X} = \{X_1, \cdots, X_M\}$ be a collection of M binary shapes, called a *family of shapes*, i.e., $X_j \in \mathcal{P}(E)$. For each shape X_j, let $\mathbf{x}_j(i) : I \rightarrow \{0, 1\}$, with $i \in I = \{1, 2, \cdots, N\}$ and $N = |E|$, be its *indicator vector*:

$$\forall X_j \in \mathcal{P}(E), \ \forall p_i \in E, \ \mathbf{x}_j(i) = \begin{cases} 1 & \text{if } p_i \in X_j, \\ 0 & \text{if } p_i \in X_j^c. \end{cases} \tag{2}$$

Then the family of shapes $\mathcal{X}$ has associated a data matrix $\mathbf{V} \in \{0, 1\}^{N \times M}$, where each indicator vector corresponds to one of its columns, i.e., $\mathbf{V}_{i,j} = \mathbf{x}_j(i)$.

3.1 Sparse NMF Approximations of Binary Sets

After computing the NMF representation on data $\mathbf{V}$, for a given constant $R > 0$, an approximation to $\mathbf{V}$ is obtained. More precisely, if we denote by $\phi_k(p_i) : E \rightarrow \mathbb{R}^+$ the basis images associated to the basis matrix $\mathbf{W}$, i.e., $\phi_k(p_i) = \mathbf{W}_{i,k}$, the following image is obtained as

$$a_{X_j}(p_i) = \sum_{k=1}^{R} \phi_k(p_i)\mathbf{H}_{k,j}. \tag{3}$$

It is obvious that without any additional constrains, function $a_X(p_i)$ is not a binary image. By the boundedness property discussed above, we have $0 \leq a_X(p_i) \leq 1$. Hence, a thresholding operation at value α is required to impose a binary approximate set $\widetilde{X}_j$ to each initial shape X_j, i.e.,

$$X_j \xrightarrow{NMF} \widetilde{X}_j : \ p_i \in \widetilde{X}_j \ \text{if} \ a_X(p_i) > \alpha. \tag{4}$$

We propose to fix, for all the examples of the paper, the threshold value to $\alpha = 0.45$, in order to favor the reconstruction of X_j against its complement.

Let us consider a practical example from a binary image collection $\mathcal{X}$, composed of $M = 100$ images from the Fish shape database ($N = 400 \times 200$), see in Fig. 2 a few examples of images. Fig. 3 depicts the corresponding basis images for various NMF algorithms: we have fixed

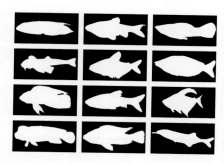

Figure 2 Examples of original binary images from the fish database, composed of 100 images.

$R = 10$ for all the cases (factor 10 of dimensionality reduction). We observe that standard NMF produces a partial part-based representation, which also includes almost complete objects for outlier shapes (basis 2–upper-center and 5-center-center). As expected, LNMF produces more local decompositions, however the orthogonality constraints involves also an atomization of some common parts. A similar problem arises for Sparse-NMF when $S_w \neq 0$ (constraint of sparsity in basis matrix $\mathbf{W}$). When the sparsity constraint is limited to the coding S_h, with a typical value around 0.6, the obtained dictionary of shapes is less local, but in exchange, this constraint involves that each binary shapes is reconstructed using a limited number of atoms. The various groups of fish shapes are therefore better approximated by the latter case than using the other NMF algorithms. The comparison of Fig. 4 illustrates qualitatively the better performance of Sparse-NMF $(S_w = 0, S_h = 0.6)$ with respect to the others. We have also included in Fig. 3A the 10 first eigenimages obtained by PCA (in red the positive values and in blue the negative values); as expected, the corresponding representation does not fit with a part-based decomposition needed for morphological operators.

3.2 Sparse Max-Approximation to Binary Dilation and Erosion

Dilation and erosion. The two fundamental morphological operators are the dilation and the erosion, which are defined respectively as the operators which commute with the union and the intersection. Given a *structuring element* $B \subseteq E$, i.e., a set defined at the origin which introduces the shape/size of the operator, the *dilation of a binary image* X by B and the *erosion of a binary image* X by B are defined respectively by (Serra, 1982, 1988;

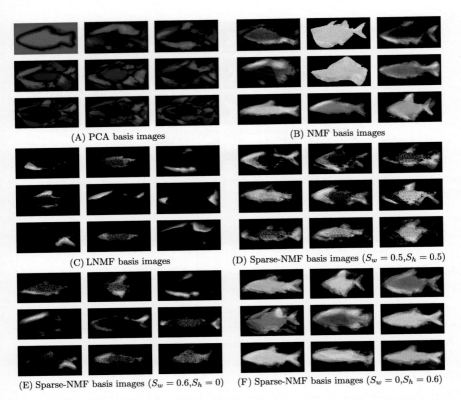

(A) PCA basis images   (B) NMF basis images

(C) LNMF basis images   (D) Sparse-NMF basis images ($S_w = 0.5, S_h = 0.5$)

(E) Sparse-NMF basis images ($S_w = 0.6, S_h = 0$)   (F) Sparse-NMF basis images ($S_w = 0, S_h = 0.6$)

Figure 3 PCA vs. non-negative representation of binary shapes. A collection of $M = 100$ shapes has been used in the experiments (see examples in Fig. 2), where the number of reduced dimensions has been fixed to $R = 10$ (in the examples are given the first 9 basis images).

Heijmans, 1994; Bloch, Heijmans, & Ronse, 2007)

$$\delta_B(X) = \cup \left\{ B(p_i) | \, p_i \in X \right\}, \tag{5}$$

and

$$\varepsilon_B(X) = \left\{ p_i \in E | \, B(p_i) \subseteq X \right\}, \tag{6}$$

where $B(p_i)$ is the structuring element centered at pixel p_i. In the case of numerical functions $f \in \mathcal{F}(E, \mathcal{T})$, which are considered in detail in next section, the *dilation of a gray-level image* is defined by (Heijmans, 1994; Soille,

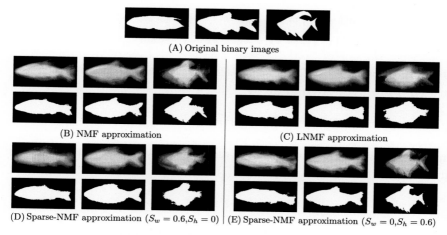

(A) Original binary images

(B) NMF approximation   (C) LNMF approximation

(D) Sparse-NMF approximation ($S_w = 0.6, S_h = 0$) | (E) Sparse-NMF approximation ($S_w = 0, S_h = 0.6$)

Figure 4 Sparse-NMF approximations to binary sets: (A) Three original shapes X_j; (B)–(E) Top, reconstructed function a_{X_j}; and Bottom, approximate set $\widetilde{X}_j$.

1999; Bloch et al., 2007)

$$\delta_B(f)(p_i) = \left\{ f(p_m) \mid f(p_m) = \sup \left[f(p_n) \right], p_n \in \check{B}(p_i) \right\} = \sup_{p_n \in \check{B}(p_i)} \left\{ f(p_n) \right\}, \quad (7)$$

and the dual *gray-level erosion* is given by (Heijmans, 1994; Soille, 1999; Bloch et al., 2007)

$$\varepsilon_B(f)(p_i) = \left\{ f(p_m) \mid f(p_m) = \inf \left[f(p_n) \right], p_n \in B(p_i) \right\} = \inf_{p_n \in B(p_i)} \left\{ f(p_n) \right\}, \quad (8)$$

where $\check{B}(p_i)$ is the transposed structuring element centered at pixel p_i. If B is symmetric with respect to the origin, one has $\check{B} = B$.

Sparse max-approximation to binary dilation. Let us first introduce two basic notions. The *indicator function* of set X, denoted $\mathbb{1}_X : E \to \{0, 1\}$, is defined by

$$\forall X \in \mathcal{P}(E), \ \forall p_i \in E, \ \mathbb{1}_X(p_i) = \begin{cases} 1 & \text{if } p_i \in X, \\ 0 & \text{if } p_i \in X^c. \end{cases} \quad (9)$$

Obviously, we have $\mathbb{1}_{X^c} = 1 - \mathbb{1}_X$. Given two sets $X, Y \in \mathcal{P}(E)$, one has the following two basic properties of indicator function:

$$
\begin{aligned}
\mathbb{1}_{X \cap Y} &= \min\{\mathbb{1}_X, \mathbb{1}_Y\} = \mathbb{1}_X \cdot \mathbb{1}_Y; \\
\mathbb{1}_{X \cup Y} &= \max\{\mathbb{1}_X, \mathbb{1}_Y\} = \mathbb{1}_X + \mathbb{1}_Y - \mathbb{1}_X \cdot \mathbb{1}_Y = \min\{1, \mathbb{1}_X + \mathbb{1}_Y\}.
\end{aligned}
$$

For a function $f : E \to \mathcal{T}$, the *thresholded set* at value $t \in \mathcal{T}$ is a mapping from $\mathcal{F}(E, \mathcal{T})$ to $\mathcal{P}(E)$ given by (Serra, 1982, 1988)

$$
\varpi_t(f) = \{ p_i \in E \,|\, f(p_i) \geq t \}. \tag{10}
$$

Using these transformations it is obvious that the binary dilation (6) can be computed using the numerical operator (7), i.e.,

$$
\delta_B(X) = \varpi_1 \left(\delta_B(\mathbb{1}_X)(p_i) \right). \tag{11}
$$

By the fundamental property of dilation (Serra, 1982, 1988), given a set defined as the union of a family of sets, i.e., $X = \cup_{k \in K} X_k$, its corresponding binary dilation by B is

$$
\delta_B(X) = \delta_B \left(\cup_{k \in K} X_k \right) = \cup_{k \in K} \delta_B(X_k). \tag{12}
$$

From (11) and the property of the indicator function for the union of sets, it is easy to see that (12) can be rewritten using the dilation of functions as

$$
\begin{aligned}
\delta_B(X) &= \varpi_1 \left(\delta_B \left(\min \left\{ 1, \sum_{k \in K} \mathbb{1}_{X_k}(p_i) \right\} \right) \right) \\
&= \varpi_1 \left(\min \left\{ 1, \sum_{k \in K} \delta_B \left(\mathbb{1}_{X_k}(p_i) \right) \right\} \right),
\end{aligned}
$$

and finally it is obtained that

$$
\delta_B(X) = \varpi_1 \left(\sum_{k \in K} \delta_B \left(\mathbb{1}_{X_k}(p_i) \right) \right). \tag{13}
$$

Hence, the justification for using NMF–based part decompositions in sparse mathematical morphology arises from Eqs. (12) and (13).

Coming back to the NMF reconstruction, expressions (3) and (4), we can now write

$$
X_j \approx \tilde{X}_j = \varpi_\alpha \left(\sum_{k=1}^{R} \phi_k(p_i) \mathbf{H}_{k,j} \right). \tag{14}
$$

Hence, we introduce the following nonlinear operator, named *sparse max-approximation to binary dilation*:

$$D_B(X_j) = \varpi_\alpha \left(\sum_{k=1}^{R} \delta_B(\phi_k)(p_i)\mathbf{H}_{k,j} \right). \tag{15}$$

Note that by the nonnegativity of $\mathbf{H}_{k,j}$, we have $\delta_B\left(\phi_k(p_i)\mathbf{H}_{k,j}\right) = \delta_B(\phi_k)(p_i)\mathbf{H}_{k,j}$. We can say that

$$\text{if } X_j \approx \widetilde{X}_j \text{ then } \delta_B(X_j) \approx D_B(X_j),$$

or in other words, the degree of approximation to the dilation depends on the degree of approximation to the original set. However neither the increasiness nor the extensivity of $D_B(X_j)$ w.r.t. X_j can be guaranteed and consequently, operator (15) is not a morphological dilation in an algebraic sense (Heijmans & Ronse, 1990).

In conclusion, in order to approximate the dilation by B of any of the M sets X_j, we only need to calculate the dilation of the R basis images. In addition, if sparsity is imposed to $\mathbf{H}$, that involves that only a limited number of dilated atoms are required for each X_j.

Dual sparse max-approximation to binary erosion. One of the most interesting properties of mathematical morphology is the duality by the complement of pairs of operators, and in particular the duality between the dilation and the erosion. Hence, the binary erosion of set X by structuring element B can be defined as the dual operator to the dilation: $\varepsilon_{\check{B}}(X) = (\delta_B(X^c))^c$, where the complement set is $X^c = \complement X = E \setminus X$. Using this property, we propose to define the *sparse max-approximation to binary erosion* as

$$E_B(X_j) = \varpi_\alpha \left(\sum_{k=1}^{R} \complement\left[\delta_{\check{B}}\left(\complement[\phi_k]\right)(p_i)\right]\mathbf{H}_{k,j} \right), \tag{16}$$

where the complement basis images are defined by $\complement[\phi_k(p_i)] = \max(\mathbf{W}_{i,k}) - \phi_k(p_i) + \min(\mathbf{W}_{i,k})$.

The results of sparse max-approximations to $D_B(X_j)$ and $E_B(X_j)$ for three examples of the Fish shapes, compared to the exact binary dilation and erosion, are given in Fig. 5. We have compared in particular the sparse max-approximation for the standard NMF and for the Sparse-NMF.

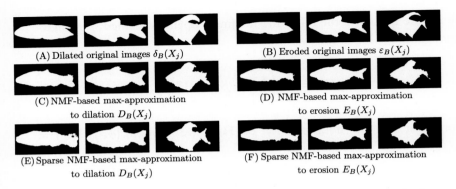

(A) Dilated original images $\delta_B(X_j)$ (B) Eroded original images $\varepsilon_B(X_j)$

(C) NMF-based max-approximation   (D) NMF-based max-approximation
to dilation $D_B(X_j)$ to erosion $E_B(X_j)$

(E) Sparse NMF-based max-approximation   (F) Sparse NMF-based max-approximation
to dilation $D_B(X_j)$ to erosion $E_B(X_j)$

Figure 5 Comparison of dilation/erosion (A)/(B) vs. sparse pseudo-operators for three examples of the Fish shapes. It is compared in particular the sparse max-approximation to dilation/erosion for the standard NMF (C)/(D) and for the Sparse-NMF (E)/(F), with ($S_w = 0$, $S_h = 0.6$). The structuring element B is a square of 5×5 pixels.

3.3 Sparse Max-Approximation to Binary Opening and Closing

Composition of dilation and erosion produces two other operators, called opening and closing, which are the fundamental bricks for morphological filtering. More precisely, opening of a binary image X by B and the dual closing of a binary image X by B are defined as (Serra, 1982, 1988; Heijmans, 1994)

$$\gamma_B(X) = \delta_B\left(\varepsilon_B(X)\right) = \bigcup \left\{ B(p_i) \mid p_i \in E \text{ and } B(p_i) \subseteq X \right\}, \qquad (17)$$

$$\varphi_B(X) = \varepsilon_B\left(\delta_B(X)\right) = \left(\gamma_B\left(X^c\right)\right)^c. \qquad (18)$$

Similarly for the case of numerical functions $f \in \mathcal{F}(E, \mathcal{T})$, opening and closing of a gray-level image by B and its dual gray-level closing are respectively given by (Serra, 1982, 1988; Heijmans, 1994)

$$\gamma_B(f)(p_i) = \delta_B\left(\varepsilon_B(f)\right)(p_i) = \sup_{p_m \in \check{B}(p_n)} \inf_{p_n \in B(p_i)} \left\{ f(p_m) \right\},$$

$$\varphi_B(f)(p_i) = \varepsilon_B\left(\delta_B(f)\right)(p_i) = \inf_{p_m \in B(p_n)} \sup_{p_n \in \check{B}(p_i)} \left\{ f(p_m) \right\}.$$

In order to apply the non-negative decomposition underlaying the NMF coding, we first introduce sufficient conditions for a compatibility of opening with set decomposition, which are based on the geometric formulation of binary opening provided by (17). Let $X = \bigcup_{k \in K} X_k$ such that

- either $\partial X_i \cap \partial X_j = \emptyset$, $\forall i, j \in K$,
- or $X_i \subseteq X_j$ or $X_j \subseteq X_i$, $\forall i, j \in K$.

Then

$$\gamma_B(X) = \gamma_B \left(\bigcup_{k \in K} X_k \right) = \bigcup_{k \in K} \gamma_B(X_k). \qquad (19)$$

Hence, opening commutes with the union only in the cases where the subsets are either totally disjoint or totally contained. We call these cases as separable subsets.

Under the assumption of separability of the atoms of the dictionary $\{\phi_k\}_{1 \leq k \leq R}$, by combining (17) and (19) (equivalence only valid on the separable case), i.e.,

$$\gamma_B(X) = \bigcup_{k \in K} \bigcup \{ B(p_i) | \, p_i \in E \text{ and } B(p_i) \subseteq X_k \}$$

$$= \bigcup \left\{ B(p_i) | \, p_i \in E \text{ and } B(p_i) \subseteq \bigcup_{k \in K} X_k \right\},$$

and similarly to the dilation and erosion, we introduce the *sparse max-approximation to binary opening and to binary closing* respectively as

$$G_B(X_j) = \varpi_\alpha \left(\sum_{k=1}^{R} \gamma_B (\phi_k) (x_i) \mathbf{H}_{k,j} \right), \qquad (20)$$

$$F_B(X_j) = \varpi_\alpha \left(\sum_{k=1}^{R} \mathsf{C} \left[\gamma_B \left(\mathsf{C}[\phi_k] \right) (p_i) \right] \mathbf{H}_{k,j} \right). \qquad (21)$$

As discussed below, in Section 4, even in the case when the separability is not always fulfilled, operators $G_B(X_j)$ and $F_B(X_j)$ are interesting to approximate the effect of the opening $\gamma_B(X_j)$ and the closing $\varphi_B(X_j)$.

3.4 Consistency and Noise Robustness of Sparse Morphological Operators

Let us discuss the following two important properties which have been empirically observed. From our viewpoint, they prove the pertinence of the sparse max–approximation to binary dilation based on Sparse–NMF representations.

Consistency. Behind this notion of consistency between Sparse–NMF and morphological binary dilation, we mean the fact that the Sparse–NMF

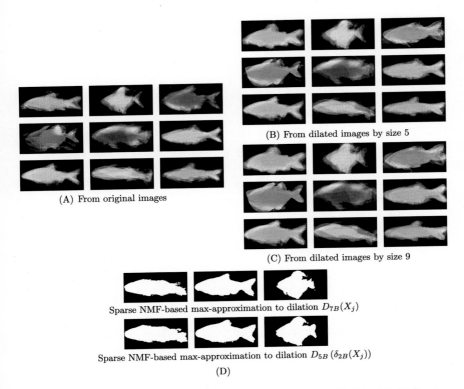

(B) From dilated images by size 5

(A) From original images

(C) From dilated images by size 9

Sparse NMF-based max-approximation to dilation $D_{7B}(X_j)$

Sparse NMF-based max-approximation to dilation $D_{5B}(\delta_{2B}(X_j))$

(D)

Figure 6 Consistency of Sparse-NMF representation for binary dilation: (A)–(C) Sparse-NMF basis images ($S_w = 0$, $S_h = 0.6$), $M = 100$ and $R = 10$; (D) max-approximation to dilation on original image collection vs. on dilated image collection.

basis from an image collection $\mathcal{X}$ should be stable to dilation of $\mathcal{X}$, i.e., $\delta_B(\mathcal{X}) = \{\delta_B(X_1), \cdots, \delta_B(X_M)\}$. Fig. 6 illustrates the consistency: it is compared in (A) the Sparse–NMF basis from the original image data set $\mathcal{X}$ to the Sparse–NMF basis obtained using the same parameters from two dilated image data set $\delta_B(\mathcal{X})$, in (B) for B a square of 5×5 and in (C) for a square 7×7. We have noted that, for a symmetric structuring element B of size relatively small with respect to the size of the objects, i.e., $|X_j \setminus \delta_B(X_j)|/|X_j| \ll 1$, $\forall j$, the consistency involves that $\sum_i \|\phi_k^{\oplus B} - \delta_B(\phi_k)\|(p_i) / \sum_i \phi_k(p_i) \approx 0$, $\forall k$, where

$$\mathcal{X} \xrightarrow{\text{Sparse-NMF}} \left\{\{\phi_k\}_{1 \le k \le R}; \mathbf{H}\right\} \Rightarrow \delta_B(\mathcal{X}) \xrightarrow{\text{Sparse-NMF}} \left\{\{\phi_k^{\oplus B}\}_{1 \le k \le R}; \mathbf{H}\right\}.$$

Using the semi-group property of multi-scale dilation δ_{nB} by homothetic convex structuring elements, i.e., $\delta_{nB} \circ \delta_{nB} = \delta_{(n+m)B}$, a consequence of such consistency is depicted in the example of Fig. 6D: the sparse max-approximation to dilation of size 7 from $\mathcal{X}$ is close the sparse max-approximation to dilation of size 5 from a Sparse-NMF representation on $\delta_{2B}(\mathcal{X})$, i.e., $D_{5B}\left(\delta_{2B}(X_j)\right) \approx D_{7B}(X_j)$. More generally, given a convex symmetric structuring element B, if one has $|X_j \setminus \delta_{mB}(X_j)|/|X_j| \ll 1$ for a scale m, then we have

$$D_{(n+m)B}(X_j) \approx \varpi_\alpha \left(\sum_{k=1}^{R} \delta_{nB}\left(\phi_k^{\oplus mB}\right)(p_i)\mathbf{H}_{k,j} \right).$$

Robustness against noise. The perturbation associated to $\mathcal{X}$ is now related to the presence of noise; in particular, since we are dealing with binary images, salt-and-pepper noise is considered. Accordingly, a given percentage p of pixels is corrupted in our dataset to obtain $\mathcal{X}^{\text{noise}-p}$. Concerning the noise, it is well known that morphological operators are very sensitive to noise and even a small amount of salt-and-pepper provides a strong perturbation of dilation/erosion. As it is illustrated in Fig. 7, Sparse-NMF dictionary learning produces atoms or basis which are quite robust against this kind of noise. If we denote by $\phi_k^{\text{noise}-p}$ the basis obtained from noisy data $\mathcal{X}'_p$, we have observed that for p up to 20–30%, we get $\sum_i \|\phi_k^{\text{noise}-p} - \phi_k\|(p_i)/\sum_i \phi_k(p_i) \approx 0$, $\forall k$, and therefore we obtain $D_B\left(X_j^{\text{noise}-p}\right) \approx D_B(X_j)$. Fig. 7D shows examples of the robustness of sparse max-approximation to dilation for $p = 10\%$ and $p = 20\%$. In conclusion, Sparse-NMF representation allows us (pseudo-)morphological processing noisy images.

4. SPARSE APPROXIMATION TO NUMERICAL MORPHOLOGICAL OPERATORS

Approaches introduced in Section 3 are extended here to gray-level images. In this case, we deal with families of discrete gray-level images, i.e., $\mathcal{F} = \{f_1(p_i), \cdots, f_M(p_i)\}$, with $f_j(p_i) \in \mathcal{F}(E, \mathcal{T})$, $\mathcal{T} = \{t_1, t_2, \cdots, t_L\}$ with $(t_{l+1} - t_l) = \Delta t$. Fig. 8 provides some examples of ORL face database (Guo, Li Stan, & Kapluk, 2000) which is used to illustrate and compare our techniques. We consider in particular two alternative paradigms: (i) each gray-scale function f_j is represented as a stack of upper level sets and Sparse-NMF processing is applied on the upper level sets, followed by image recomposition from processed upper level sets; (ii) straightforward Sparse-NMF representation and processing of gray-scale images.

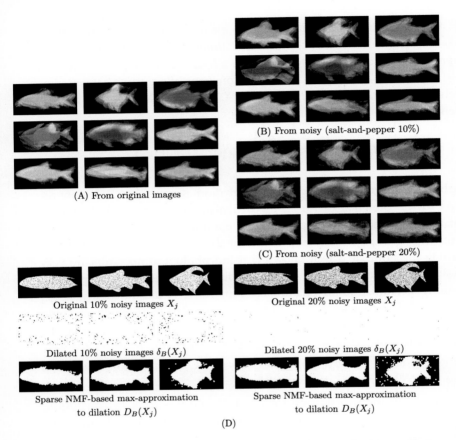

(A) From original images

(B) From noisy (salt-and-pepper 10%)

(C) From noisy (salt-and-pepper 20%)

Original 10% noisy images X_j

Dilated 10% noisy images $\delta_B(X_j)$

Sparse NMF-based max-approximation
to dilation $D_B(X_j)$

Original 20% noisy images X_j

Dilated 20% noisy images $\delta_B(X_j)$

Sparse NMF-based max-approximation
to dilation $D_B(X_j)$

(D)

Figure 7 Robustness against noise of Sparse-NMF representation for binary dilation: (A)–(C) Sparse-NMF basis images ($S_w = 0$, $S_h = 0.6$), $M = 100$ and $R = 10$; (D) noisy binary images, dilated noisy images and max-approximation to dilation of noisy images.

4.1 Sparse-NMF Processing of Upper Level Sets

The thresholded set of f_j at each t_l, i.e., $X_j^{t_l} = \varpi_{t_l}(f_j)$, is called the upper level-set at t_l of f_j. The set of upper level sets constitutes a family of decreasing sets (Serra, 1982, 1988):

$$t_\lambda \geq t_\mu \Rightarrow X^{t_\lambda} \subseteq X^{t_\mu} \quad \text{and} \quad X^{t_\lambda} = \cap\{X^{t_\mu}, \mu < \lambda\}.$$

Moreover, any (semi–continuous) image f_j can be viewed as a unique stack of its upper level sets, which leads to the following reconstruction property

Figure 8 Examples of original numerical images from the ORL face database.

(Serra, 1982, 1988):

$$f_j(p_i) = \sup\{t_l|\, p_i \in X_j^{t_l}\}, \quad t_l \in \mathcal{T}.$$

We prefer here to consider the alternative reconstruction using a numerical sum of the indicator function of upper level sets (Wendt, Coyle, & Gallagher, 1986; Ronse, 2009):

$$f_j(p_i) = \Delta t \sum_{l=1}^{L} \mathbb{1}_{X_j^{t_l}}(p_i). \tag{22}$$

It is well known in mathematical morphology that any binary increasing operator, such as the dilation and erosion, can be generalized to gray-level images by applying the binary operator to each cross-section, and then by reconstructing the corresponding gray-level image (Serra, 1982, 1988; Ronse, 2009), i.e.,

$$\delta_B(f_j)(p_i) = \Delta t \sum_{l=1}^{L} \mathbb{1}_{\delta_B\left(X_j^{t_l}\right)}(p_i), \tag{23}$$

$$\varepsilon_B(f_j)(p_i) = \Delta t \sum_{l=1}^{L} \mathbb{1}_{\varepsilon_B\left(X_j^{t_l}\right)}(p_i). \tag{24}$$

Consider now that each image of the initial gray-level family $\mathcal{F}$ of M images is decomposed into its L upper level set. Hence, we have

$$\mathcal{F} = \{f_1, \cdots, f_M\} \mapsto \mathcal{X} = \{X_1^{t_1}, X_1^{t_2} \cdots, X_1^{t_L}, \cdots, X_{M-1}^{t_L}, X_M^{t_1}, \cdots, X_M^{t_L}\},$$

where $\mathcal{X}$ is a family of $M' = M \times L$ binary images. Therefore, we can use NMF algorithms, for a given dimension R, to approximate each set $X_j^{t_l}$ and then approximate the corresponding function $f_j(p_i)$. Thus, using the results of the previous section, we are able now to introduce the following definition for the *sparse max-approximation to gray-level dilation and erosion* given respectively by

$$D_B(f_j)(p_i) = \Delta t \sum_{l=1}^{L} \mathbb{1}_{D_B(X_j^{t_l})}(p_i), \tag{25}$$

and

$$E_B(f_j)(p_i) = \Delta t \sum_{l=1}^{L} \mathbb{1}_{E_B(X_j^{t_l})}(p_i), \tag{26}$$

with

$$D_B\left(X_j^{t_l}\right) = \varpi_\alpha \left(\sum_{k=1}^{R} \delta\left(\phi_k\right)(p_i) \mathbf{H}_{k,j+l} \right),$$

$$E_B\left(X_j^{t_l}\right) = \varpi_\alpha \left(\sum_{k=1}^{R} \mathsf{C}\left[\delta_B\left(\mathsf{C}[\phi_k]\right)(p_i)\right] \mathbf{H}_{k,j+l} \right).$$

Similarly, a sparse max-approximation to gray-scale opening and closing can be achieved by means of the expressions (20) and (21) for respectively $G_B(X_j)$ and $F_B(X_j)$ applied to the upper level sets of f_j.

Behavior of sparse max-approximation to gray-level dilation and erosion based on upper level set decomposition is shown in Fig. 9. Note that, for this example, the number of initial images $M = 20$, and each of them has been quantized in $L = 10$ gray levels, i.e., $M' = 20 \times 10 = 200$ dimensions and them for Sparse-NMF the number of atoms is fixed to $R = 75$. Hence a reduction factor which is not significant. Some of the corresponding atoms are given in Fig. 9A. As expected the quality of the sparse max-approximation to dilation and to erosion depends on the quality of the initial Sparse-NMF reconstruction of the image. For instance, the first face (man with glasses) is not well approximated with the learned NMF basis

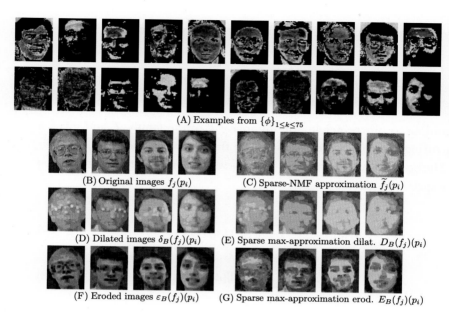

(A) Examples from $\{\phi\}_{1 \leq k \leq 75}$

(B) Original images $f_j(p_i)$ (C) Sparse-NMF approximation $\tilde{f}_j(p_i)$

(D) Dilated images $\delta_B(f_j)(p_i)$ (E) Sparse max-approximation dilat. $D_B(f_j)(p_i)$

(F) Eroded images $\varepsilon_B(f_j)(p_i)$ (G) Sparse max-approximation erod. $E_B(f_j)(p_i)$

Figure 9 A collection of $M = 20$ faces has been used in this experiment, where the number of reduced dimensions for the binary matrix **V** has been fixed to $R = 75$ (note that $M' = 20 \times 10 = 200$ dimensions). Top, some examples of the Sparse-NMF atoms obtained from the 200 upper level sets. Bottom, four examples of the ORL face database (B) (quantized in $L = 10$ gray levels) and (C) corresponding approximated images using Sparse-NMF. Comparison of dilation/erosion (D)/(F) vs. sparse max-approximation to dilation/erosion for the Sparse-NMF (E)/(G). The structuring element B is a square of size 3×3 pixels.

and hence, its approximated dilation and erosion are also unsatisfactory. On the contrary, in the case of the last image (woman), the results are more relevant. Nevertheless, at this point, we can conclude that the sparse approximation using upper level set stack, which is theoretically consistent with the binary framework discussed in Section 3 is not very useful in practice.

4.2 Sparse-NMF Representation and Processing of Gray-Scale Images

As an alternative to the previous formulation, we propose a straightforward use of Sparse–NMF representation from the image gray-scale image collection $\mathcal{F} = \{f_1(p_i), \cdots, f_M(p_i)\}$. That is, each column of the data matrix **V** corresponds to an image, i.e., $V_{i,j} = f_j(p_i)$. Then, after Sparse–NMF dictionary learning of dimension R, each image is approximated as

$$f_j(p_i) \xrightarrow{\text{Sparse-NMF}} \tilde{f_j}(p_i) = \sum_{k=1}^{R} \phi_k(p_i)\mathbf{H}_{k,j}.$$

According to our principle, for any image f_j of the family $\mathcal{F}$ and given a structuring element B, we can introduce the following operators.

The *sparse max-approximation to gray-level dilation and to gray-level erosion* are respectively given by

$$D_B(f_j)(x_i) = \sum_{k=1}^{R} \delta_B\left(\phi_k\right)(x_i)\mathbf{H}_{k,j}, \tag{27}$$

$$E_B(f_j)(x_i) = \sum_{k=1}^{R} \complement\left[\delta_B\left(\complement[\phi_k]\right)(x_i)\right]\mathbf{H}_{k,j}. \tag{28}$$

In addition, the *sparse max-approximation to gray-level opening and gray-level closing* are respectively defined as

$$G_B(f_j)(x_i) = \sum_{k=1}^{R} \gamma_B\left(\phi_k\right)(x_i)\mathbf{H}_{k,j}, \tag{29}$$

$$F_B(f_j)(x_i) = \sum_{k=1}^{R} \complement\left[\gamma_B\left(\complement[\phi_k]\right)(x_i)\right]\mathbf{H}_{k,j}. \tag{30}$$

We must point out again that these approximate nonlinear operators do not satisfy the standard properties of gray-level dilation and erosion.

Similarly to the binary case, let us discuss the conditions under which these max-approximations to dilation and opening are more relevant. First, let us rewrite (27) as supremum of cylinders (Heijmans, 1994; Bloch et al., 2007):

$$\delta_B(f) = \bigvee \left\{ C_{B(p_i),\,f(p_i)} \,:\, p_i \in E \right\},$$

where $C_{B,t}$ is the cylinder of base B at the origin and height t. From this expression, it is easy to see that dilation commutes with supremum of functions, i.e.,

$$\delta_B \left(\bigvee_{k \in K} f_k \right) = \bigvee \left\{ C_{B(p_i),\, \bigvee_{k \in K} f_k(p_i)} \,:\, p_i \in E \right\}$$

$$= \bigvee \left\{ \bigvee_{k \in K} C_{B(p_i),\, f_k(p_i)} \,:\, p_i \in E \right\}$$

$$= \bigvee_{k \in K} \bigvee \left\{ C_{B(p_i),\, f_k(p_i)} \,:\, p_i \in E \right\} = \bigvee_{k \in K} \delta_B \left(f_k \right).$$

For the case of the opening (29), one has (Heijmans, 1994; Bloch et al., 2007):

$$\gamma_B(f) = \bigvee \left\{ C_{B(p_i),\, t} \,:\, p_i \in E,\ C_{B(p_i),\, t} \leq f \right\}.$$

Thus, a sufficient condition for the commutation of opening with the supremum is the "disjointness" of functions, i.e., $\bigwedge_{k \in K} f_k = 0$. It is obvious also that for "disjoint functions", one has: $\bigvee_{k \in K} f_k = \sum_{k \in K} f_k$. Therefore, operators (27) and (29) will provide a good approximation to dilation and opening in the case where the Sparse-NMF representation is composed of disjoint atoms, i.e.,

$$\bigwedge \phi_k(p_i) \approx 0,$$

which turns out to be a condition of non-redundancy of the Sparse-NMF basis.

Fig. 10A gives the Sparse-NMF representation of the faces, where a collection of $M = 80$ images has been used in this experiment and dimension of dictionary has been fixed to $R = 20$ atoms. As we can observe, the effectivity of the max-approximation to dilation and opening is consistent with the approximation to the corresponding original image. Thus, such approximations can be then used to compute evolved morphological operators. We have in particular illustrated in Fig. 10 two operators which can be applied in face image feature extraction. The first one is the morphological gradient, i.e., $\delta(B)(f) - \varepsilon(B)(f)$, typically used for contour detection. The second one, the black top-hat, i.e., $\varphi(B)(f) - f$, extracts the dark structures, which in the present case of faces, correspond just to the eyes and lips.

5. APPLICATIONS

We consider in this section to potential applications of Sparse-NMF morphological processing to problems arising from multi/hyperspectral image processing.

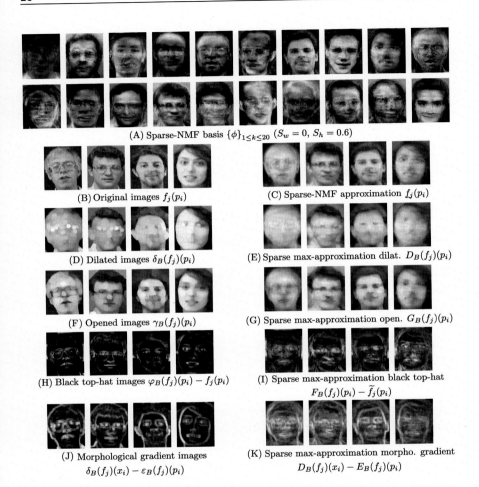

(A) Sparse-NMF basis $\{\phi\}_{1 \leq k \leq 20}$ $(S_w = 0,\, S_h = 0.6)$

(B) Original images $f_j(p_i)$

(C) Sparse-NMF approximation $f_j(p_i)$

(D) Dilated images $\delta_B(f_j)(p_i)$

(E) Sparse max-approximation dilat. $D_B(f_j)(p_i)$

(F) Opened images $\gamma_B(f_j)(p_i)$

(G) Sparse max-approximation open. $G_B(f_j)(p_i)$

(H) Black top-hat images $\varphi_B(f_j)(p_i) - f_j(p_i)$

(I) Sparse max-approximation black top-hat $F_B(f_j)(p_i) - \widetilde{f}_j(p_i)$

(J) Morphological gradient images $\delta_B(f_j)(x_i) - \varepsilon_B(f_j)(p_i)$

(K) Sparse max-approximation morpho. gradient $D_B(f_j)(x_i) - E_B(f_j)(p_i)$

Figure 10 Sparse-NMF representation and processing of gray-scale images $\mathcal{F}$: A collection of $M = 80$ faces has been used in this experiment and dimension of dictionary has been fixed to $R = 20$. Top, the obtained Sparse-NMF atoms. Bottom, four examples of the ORL face database (B) and (C) corresponding approximated image using Sparse-NMF. Comparison of dilation, opening, black top-hat and gradient (D)/(F)/(H)/(J) vs. sparse max-approximation to this operators. The structuring element B is a square of size 5×5 pixels for the dilation and opening, size 9×9 for the black top-hat.

5.1 Sparse Processing of Multivariate Boolean Textures

The motivation of this study is to deal with random image models on multivalued images, and more precisely with the case of multivariate Boolean random set model (Jeulin, 1991). This approach is an extension of the classi-

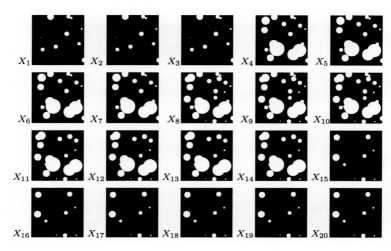

Figure 11 Example of 2D multivariate image of 20 components $\mathbf{X} = (X_1, X_2, \cdots, X_{20})$, $X_i \in \mathcal{P}(E)$, $E = [0, 255] \times [0, 255]$, $\theta|E| = 30$, A_i are random disks.

cal Boolean random model (Matheron, 1975). Multivariate texture images which fit with this model appear in different fields: multi/hyperspectral images, energy dispersive spectrometry, electron energy-loss spectrometry, etc.

Let us consider a Poisson point process in a set $E \subset \mathbb{R}^n$ with intensity (average number of point per unit of volume) θ. In each point $x_k \in E$ of the process, an independent realization of a multivariate random compact set $\mathbf{A} = (A_1, A_2, \cdots, A_d)$ is implanted: a realization of the primary grain sets A_i is placed at x_k. Each component X_i of the multivariate Boolean random set $\mathbf{X} = (X_1, X_2, \cdots, X_d)$, $\mathbf{X} \sim (\theta, \mathbf{A})$ is a Boolean random closed set obtained as $X_i = \bigcup_k A_i(x_k)$. An example of 2D multivariate image of 20 components $\mathbf{X} = (X_1, X_2, \cdots, X_{20})$, $X_i \in \mathcal{P}(E)$, $E = [0, 255] \times [0, 255]$, $\theta|E| = 30$, and where A_i are random disks is given in Fig. 11. In practical situations, this kind of binary textures can be for instance obtained after segmentation of multi/hyperspectral images.

In the framework of the theory of random sets, these structures and their models are fully characterized by a functional called Choquet capacity (Matheron, 1975). In the present model, the multivariate Choquet capacity $T_{\mathbf{X}}(\mathbf{K})$, with $\mathbf{K} = (K_1, K_2, \cdots, K_d)$, $K_i \in \mathcal{K}$ (compact set), is defined as (Jeulin, 1991)

$$T_{\mathbf{X}}(\mathbf{K}) = 1 - Q_{\mathbf{X}}(\mathbf{K}) = 1 - \exp\left(-\theta\overline{\mu}_n\left(\mathbf{A} \oplus_\cup \check{\mathbf{K}}\right)\right),$$

with

$$\mathbf{A} \oplus_{\cup} \check{\mathbf{K}} = \left\{ A_1 \oplus \check{K}_1 \cup A_2 \oplus \check{K}_2 \cup \cdots A_d \oplus \check{K}_d \right\},$$

$\bar{\mu}_n$ being the Lebesgue measure in $\mathbb{R}^n$ and where $X \oplus \check{K}$ is the dilation of set $X \in \mathcal{P}(E)$ by structuring element K, i.e.,

$$\delta_K(X) = X \oplus \check{K} = \bigcup_{p_i \in \check{K}} X(p_i),$$

where $\check{K} = \{-p_i : p_i \in K\}$. Hence, an experimental estimation of $T_{\mathbf{X}}(\mathbf{K})$ can be obtained by morphological dilations on realizations of components of $\mathbf{K}$. If a single grain is used for each Poisson point, the components are independent Boolean random sets (Matheron, 1975; Jeulin, 1991); otherwise some correlations between components are present.

As a fundamental property, this model is stable by the union; such that any number of components (even correlated) of a Boolean multivariate random set is still a Boolean random set. Hence a notion of dimensionally reduction by supremum of primary schemes is compatible with the model. In our current example of Fig. 11, the A_i random disks have been simulated using three vector prototypes, see Fig. 12B which consequently involves correlations between the components.

Using our theory of Sparse-NMF representation and max-approximation to dilation we note that, on the one hand, the Sparse-NMF basis $\{\phi_k(p_j)\}$ gives just a dimensionality reduction of the initial multivariate Boolean model. In other words, thresholding the four components of Fig. 12A produces a Boolean model of dimension 4 instead of dimension 20 of the original model. We observe in the coding vectors $\mathbf{H}$ associated to basis, Fig. 12C, that the four components are consistent with the decomposition of the original spectra Fig. 12D. On the other hand, approximated dilations to the initial components of the model can be naturally obtained using expression (15), Fig. 12D–F.

In summary, in order to model a multispectral Boolean set $\mathbf{X}$, we need to compute the characteristic curves related to Choquet capacity $T_{\mathbf{X}}(\mathbf{K})$, for different $\mathbf{K}$, and these computations can be efficiently approximated in a consistent theoretical way with our Sparse-NMF dilations.

5.2 Sparse Processing of Hyperspectral Images

High-spatial resolution hyperspectral images are used nowadays in remote sensing and other application domains. They constitute high dimensional

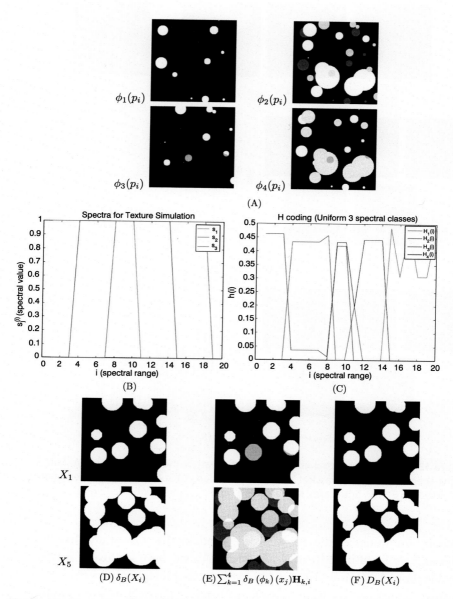

Figure 12 Dimensionality reduction of multivariate Boolean set **X** from Fig. 11: (A) Sparse-NMF basis with $R = 4$; (B) three spectra used to simulate multivariate set **X**; (C) coding vectors **H** associated to basis of (A); (D) two dilated components of **X**; (E) approximated dilation using the Sparse-NMF basis; (F) thresholded images.

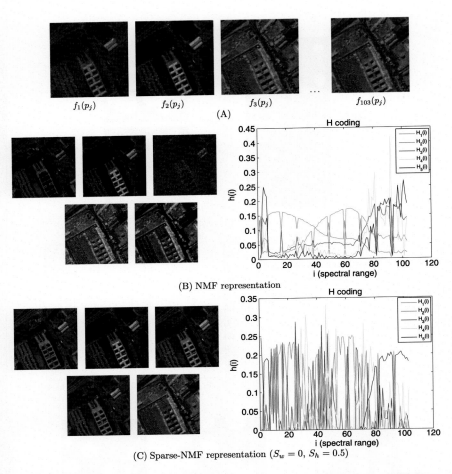

$$f_1(p_j) \qquad f_2(p_j) \qquad f_3(p_j) \qquad \cdots \qquad f_{103}(p_j)$$

(A)

(B) NMF representation

(C) Sparse-NMF representation ($S_w = 0$, $S_h = 0.5$)

Figure 13 Pavia hyperspectral images (crop of a part of the whole image): (A) Some of the spectral bands, (B) classical NMF representation, (C) Sparse-NMF representation ($S_w = 0$, $S_h = 0.5$). In both cases, left, basis $\{\phi_k(p_j)\}_{1 \leq k \leq 5}$ and right, coding **H**.

datasets: for instance, the Pavia image, used in this case-study, has dimensions $N = 340 \times 610$ and $M = 103$. As we have discussed in the introduction, NMF is widely used in hyperspectral imaging since the non-negative representation is compatible with linear physical model of end-members/abundances.

Fig. 13 shows some of the spectral bands of Pavia image in (A) as well as the classical NMF representation compared to the Sparse-NMF counterpart (with $S_w = 0$ and $S_h = 0.5$), where for both cases $R = 5$. We have

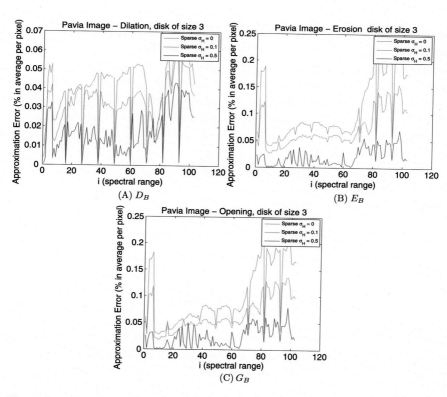

Figure 14 Approximation error (in average percentage) to morphological operators of the $M = 103$ spectral bands $\{f_i\}$, with respect to the sparseness parameter S_h (in red, $S_h = 0$; in green, $S_h = 0.1$; and in blue, $S_h = 0.5$): (A) Max-approximation to dilation D_B, (B) max-approximation to erosion E_B, (C) max-approximation to opening G_B. For all the cases B is a square 3×3.

assessed the effect of the sparseness parameter S_h on the approximation error (in average percentage) to morphological operators: max–approximation to dilation D_B (27), to erosion E_B (28), and to opening G_B (29), for the $M = 103$ spectral bands (Fig. 14). From this example we clearly observe that with $S_h = 0.5$ the approximation error is much better than without any sparseness. We note also that a low value of S_h can eventually produce worst results than unconstrained NMF.

Mathematical morphology is widely used in remote sensing hyperspectral imaging. One of the most popular applications in the state–of–the–art is the spectral–spatial classification based on the notion of morphological profile (Pesaresi & Benediktsson, 2001), and its various extensions to hyperspectral images (Fauvel, Benediktsson, Chanussot, & Sveinsson, 2008; Velasco-Forero & Angulo, 2013). Morphological profiles are founded on the granulometry (or opening-based scale-space). We remind that a granulometry (Serra, 1982, 1988) is a family of openings $\{\gamma_{B_n}\}_{0 \le n \le L}$, depending on a (discrete) positive parameter n, representing scale of homothetic structuring element, i.e., $B_n = nB$. Then, given an image f, it produces a multi-scale decomposition of bright structures:

$$s_n(f)(p_j) = f(p_j) - \gamma_{B_n}(f)(p_j),$$

or using a differential representation:

$$r_n(f)(p_j) = \gamma_{B_{n-1}}(f)(p_j) - \gamma_{B_n}(f)(p_j)$$

such that

$$f(p_j) = \gamma_{B_L}(f)(p_j) + \sum_{n=1}^{L} r_n(f)(p_j).$$

Then, this multi-scale decomposition can be used either for tensor-based spatial–spectral dimensionality reduction (Velasco-Forero & Angulo, 2013) (4D tensor = 2D space × 1D spectrum × 1D morphology):

$$\left\{ \{f_\lambda\}_{1 \le \lambda \le M}, \{\gamma_{B_1}(f_\lambda)\}_{1 \le \lambda \le M} \cdots, \{\gamma_{B_L}(f_\lambda)\}_{1 \le \lambda \le M} \right\};$$

or for spatial–spectral classification, where the feature vector per pixel p_j given by

$$\left(f_1(p_j), f_2(p_j), \cdots, f_M(p_j), \gamma_{B_1}(f_1)(p_j), \gamma_{B_1}(f_2)(p_j), \cdots, \gamma_{B_L}(f_M)(p_j) \right)$$

is the so-called morphological profile at p_j. As shown in Fig. 15 for two examples of spectral bands, the max-approximation to opening-based scale-space using Sparse-NMF representation is quite satisfactory and can be naturally used in morphological profile-based spectral spatial classification.

6. CONCLUSIONS AND PERSPECTIVES

We have introduced the notion of sparse binary and gray-level max-approximation to morphological operators based on Sparse-NMF representation.

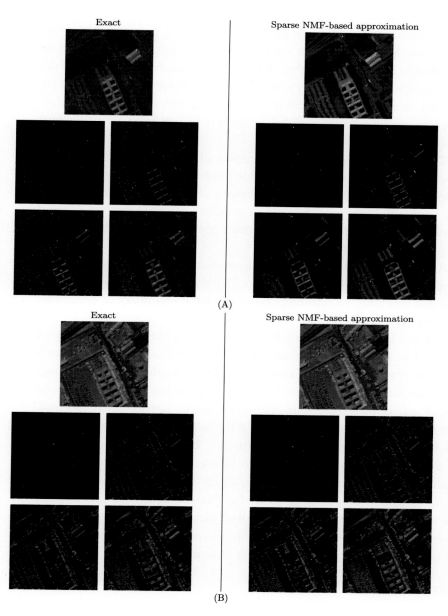

Figure 15 Comparison of opening-based scale-space $f_i(x)$, $s_1(f_i)(x)$, $s_3(f_i)(x)$, $s_5(f_i)(x)$, $s_7(f_i)(x)$ for exact openings and for Sparse-NMF max-approximations ($R = 5$, $S_h = 0.5$). In (A) for spectral band $f_1(p_j)$ and in (B) for $f_3(p_j)$.

We have linked the role of non-negative decompositions of sets of functions, viewed as vectors in a high dimensional space, to the commutation by supremum of the dilation operator, which is just the counterpart of convolution in max-plus algebra. We have also seen that in the case of the opening operator, additional conditions of separability (or disjointness) are required for an exact compatibility. Empirical examples showed that this assumption can be relaxed since the obtained approximations are satisfactory in practical cases.

We have illustrated the practical interest of our approach for morphological processing of multivariate vector images, namely multispectral/hyperspectral images. Our results are relatively encouraging and they open a new avenue to study how the current paradigm of sparse modeling (based mainly on linear operations) in computer vision can be extended to the nonlinear morphological framework.

As we have discussed, NMF produces linear non-negative decompositions which are well indicated for morphological operators in the case of separable functions, i.e., decomposition by sum is equivalent to decomposition by max.

It is well known that morphological operators are linear in the max-plus algebra. Consequently, the intrinsic linear decomposition in such algebra is the more appropriate one to introduced sparse morphological processing, without separability conditions. Some recent work has addressed the problem of matrix factorization in max-plus algebra, see for instance Hook (2014), which is based on some classical results on max-plus spectral theory and max-plus eigenvalues as the asymptotic behavior of standard eigenvalues by means of a nonlinearization of the matrix (Gaubert, Akian, & Bapat, 2001; De Schutter & De Moor, 2002). Other recent results on matrix factorization over max-times algebra (Karaev, 2013) can also be used as a representation approach for morphological operators. To explore the interest of decompositions on max-plus and max-times algebras for morphological processing will be the object of future research.

REFERENCES

Angulo, J., & Velasco-Forero, S. (2011). Sparse mathematical morphology using non-negative matrix factorization. In *Lecture notes in computer science: Vol. 6671. Proc. of 2011 international symposium on mathematical morphology (ISMM'11)* (pp. 1–12). Springer.

Arora, S., Ge, R., Kannan, R., & Moitra, A. (2012). Computing a nonnegative matrix factorization – provably. In *Proc. of the 44th symposium on theory of computing* (pp. 145–162).

Bloch, I., Heijmans, H., & Ronse, C. (2007). Mathematical morphology. In M. Aiello, I. Pratt-Hartmann, & J. van Benthem (Eds.), *Handbook of spatial logics* (pp. 857–944). Springer.

Cai, D., He, X., Han, J., & Huang, T. S. (2011). Graph regularized nonnegative matrix factorization for data representation. *IEEE Transactions on Pattern Analysis and Machine Intelligence, 33*(8), 1548–1560.

Cichocki, A., Zdunek, R., Phan, A. H., & Amari, S. I. (2009). *Nonnegative matrix and tensor factorizations*. John Wiley & Sons.

Damle, A., & Sun, Y. (2014). Random projections for non-negative matrix factorization. Retrieved from arXiv:1405.4275.

De Schutter, B., & De Moor, B. (2002). The QR decomposition and the singular value decomposition in the symmetrized max-plus algebra revisited. *SIAM Review, 44*(3), 417–454.

Ding, C., He, X., & Simon, H. D. (2005). On the equivalence of nonnegative matrix factorization and spectral clustering. In *Proc. of SIAM data mining conference (SDM'05)* (pp. 606–610).

Donoho, D. (2006). Compressed sensing. *IEEE Transactions on Information Theory, 52*(4), 1289–1306.

Donoho, D., & Stodden, V. (2004). When does non-negative matrix factorization give a correct decomposition into parts? In *Advances in neural information processing: Vol. 16. Proc. NIPS'03*. MIT Press.

Elad, M., & Aharon, M. (2006). Image denoising via sparse and redundant representations over learned dictionaries. *IEEE Transactions on Image Processing, 15*(12), 3736–3745.

Esser, E., Möller, M., Osher, S., Sapiro, G., & Xin, J. (2012). A convex model for non-negative matrix factorization and dimensionality reduction on physical space. *IEEE Transactions on Image Processing, 21*(7), 3239–3252.

Fauvel, M., Benediktsson, J. A., Chanussot, J., & Sveinsson, J. R. (2008). Spectral and spatial classification of hyperspectral data using SVMs and morphological profiles. *IEEE Transactions on Geoscience and Remote Sensing, 46*(11), 3804–3814.

Gaubert, S., Akian, M., & Bapat, R. (2001). Generic asymptotics of eigenvalues using min-plus algebra. In *Proc. of the satellite workshop on max-plus algebras, IFAC SSSC01*. Elsevier.

Guan, N., Tao, D., Luo, Z., & Yuan, B. (2011). Manifold regularized discriminative non-negative matrix factorization with fast gradient descent. *IEEE Transactions on Image Processing, 20*(7), 2030–2048.

Guo, G., Li Stan, Z., & Kapluk, C. (2000). Face recognition by support vector machines. In *Fourth IEEE international conference on automatic face and gesture recognition*.

Heijmans, H. J. A. M. (1994). *Morphological image operators*. Boston: Academic Press.

Heijmans, H. J. A. M., & Ronse, Ch. (1990). The algebraic basis of mathematical morphology I. Dilations and erosions. *Computer Vision, Graphics, and Image Processing, 50*(3), 245–295.

Hook, J. (2014). *Max-plus singular values* (MIMS EPrint 2014.7). UK: The University of Manchester.

Hoyer, P. (2004). Non-negative matrix factorization with sparseness constraints. *Journal of Machine Learning Research, 5*, 1457–1469.

Huang, K., Sidiropoulos, N., & Swami, A. (2014). Non-negative matrix factorization revisited: Uniqueness and algorithm for symmetric decomposition. *IEEE Transactions on Signal Processing, 62*(1), 211–224.

Jeulin, D. (1991). Multivariate random image models. *Acta Stereologica, 11*, 59–66.

Karaev, S. (2013). *Matrix factorization over max-times algebra for data mining* (Master's thesis, supervisor P. Miettinen). Saarbrücken, Germany: Universität des Saarlandes.

Kumar, A., Sindhwani, V., & Kambadur, P. (2013). Fast conical hull algorithms for near-separable non-negative matrix factorization. In *Proc. of the 30th international conference on machine learning*.

Lee, D. D., & Seung, H. S. (1999). Learning the parts of objects by non-negative matrix factorization. *Nature, 401*(6755), 788–791.

Lee, D. D., & Seung, H. S. (2001). Algorithms for non-negative matrix factorization. In *Advances in neural information processing: Vol. 13. Proc. NIPS'00*. MIT Press.

Li, S. Z., Hou, X., Zhang, H., & Cheng, Q. (2001). Learning spatially localized parts-based representations. In *Proc. IEEE conference on computer vision and pattern recognition (CVPR), Vol. I* (pp. 207–212).

Mairal, J., Bach, F., & Ponce, J. (2012). Sparse modeling for image and vision processing. *Foundations and Trends in Computer Graphics and Vision, 8*(2–3), 85–283.

Mairal, J., Elad, M., & Sapiro, G. (2008). Sparse representation for color image restoration. *IEEE Transactions on Image Processing, 17*(1), 53–69.

Matheron, G. (1975). *Random sets and integral geometry*. New York: Wiley.

Miettinen, P. (2010). Sparse Boolean matrix factorizations. In *Proc. of IEEE 10th international conference on data mining (ICDM'10)* (pp. 935–940).

Miettinen, P., Mielikainen, T., Gionis, A., Das, G., & Mannila, H. (2008). The discrete basis problem. *IEEE Transactions on Knowledge and Data Engineering, 20*(10), 1348–1362.

Pesaresi, M., & Benediktsson, J. A. (2001). A new approach for the morphological segmentation of high resolution satellite imagery. *IEEE Transactions on Geoscience and Remote Sensing, 39*(2), 309–320.

Plumbley, M. D. (2003). Algorithms for nonnegative independent component analysis. *IEEE Transactions on Neural Networks, 14*(3), 534–543.

Ronse, Ch. (2009). Bounded variation in posets, with applications in morphological image processing. In M. Passare (Ed.), *Acta universitatis Upsaliensis: Vol. 86. Proceedings of the Kiselmanfest-2006* (pp. 249–281).

Schachtner, R., Pöppel, G., & Lang, E. W. (2010). A nonnegative blind source separation model for binary test data. *IEEE Transactions on Circuits and Systems I, 57*(7), 1439–1448.

Serra, J. (1982). *Image analysis and mathematical morphology. Vol. I.* London: Academic Press.

Serra, J. (1988). *Image analysis and mathematical morphology. Vol. II. Theoretical advances.* London: Academic Press.

Soille, P. (1999). *Morphological image analysis*. Berlin: Springer-Verlag.

Theis, F., Stadlthanner, K., & Tanaka, T. (2005). First results on uniqueness of sparse nonnegative matrix factorization. In *Proc. of the 13th European signal processing conference (EUSIPCO'05)* (pp. 1–4).

Tomé, A. M., Schachtner, R., Vigneron, V., Puntonet, C. G., & Lang, E. W. (2015). A logistic non-negative matrix factorization approach to binary data sets. *Multidimensional Systems and Signal Processing, 26*(1), 125–143.

Vavasis, S. A. (2009). On the complexity of nonnegative matrix factorization. *SIAM Journal on Optimization, 20*, 1364–1377.

Velasco-Forero, S., & Angulo, J. (2013). Classification of hyperspectral images by tensor modeling and additive morphological decomposition. *Pattern Recognition, 46*(2), 566–577.

Wendt, P. D., Coyle, E. J., & Gallagher, N. C. (1986). Stack filters. *IEEE Transactions on Acoustics, Speech, and Signal Processing, 34*(4), 898–911.

Yu, G., Sapiro, G., & Mallat, S. (2011). Solving inverse problems with piecewise linear estimators: From Gaussian mixture models to structured sparsity. *IEEE Transactions on Image Processing, 21*(5), 2481–2499.

Yuan, Y., Li, X., Pang, Y., Lu, X., & Tao, D. (2009). Binary sparse nonnegative matrix factorization. *IEEE Transactions on Circuits and Systems for Video Technology, 19*(5), 772–777.

Zafeiriou, S., & Petrou, M. (2010). Nonlinear non-negative component analysis algorithms. *IEEE Transactions on Image Processing, 19*(4), 1050–1066.

Zass, R., & Shashua, A. (2006). Nonnegative sparse PCA. In *Advances in neural information processing systems* (pp. 1561–1568).

Zhang, Z., Ding, C., Li, T., & Zhang, X. (2007). Binary matrix factorization with applications. In *Proc. of IEEE seventh international conference on data mining (ICDM'07)* (pp. 391–400).

CHAPTER TWO

Disorder Modifications of the Critical Temperature for Superconductivity: A Perspective from the Point of View of Nanoscience

Clifford M. Krowne
Naval Research Laboratory, Washington, DC, United States
e-mail address: cliff.krowne@nrl.navy.mil

Contents

1. INTRODUCTION

Superconductivity can be modified by various effects related to randomness, disorder, structural defects, and other similar physical effects. Their affects on superconductivity are important because such effects are intrinsic to certain material system's preparation, or may be intentionally produced. In this chapter, we show in the context of a Cooper instability relationship, that introduction of disorder through impurities could

Advances in Imaging and Electron Physics, Volume 202
ISSN 1076-5670
http://dx.doi.org/10.1016/bs.aiep.2017.07.002

39

possibly lead to an increase in T_c. This is an old subject, having been addressed decades ago in the view of simpler substances, including metal alloy materials. Today, with the advances in material science, nanoscience, and atomic level preparation of materials and devices, this subject should be reexamined. That is our purpose here, especially in light of some recent discoveries made in the area of the metal–insulator transition, to be covered below shortly. Although there is a complicated collection of quantum Green's functions, polarization functions (correlation functions), and vertex functions, used in the hierarchy of assumptions to extract out an analytical formula for T_c, one may obtain a formula for it which is recognizable, and relatable to BCS theory, by using many-body quantum field theory for condensed matter expeditiously. Handling modified BCS superconductors, medium temperature superconductors, and high T_c superconductors may be possible by appropriate use of more accurate function representations and appropriate adaptions. All of this analysis is done while not necessarily maintaining band structure symmetry at various levels of development. The reader will find use of perturbational Feynman diagrammatic techniques, finite temperature quantum Matsubara Green's functions, and quantum perturbational derivations without the diagrams.

Superconductivity may be able to play a new or improved role in advancing technologies such as wireless communications, satellites, and other electronic and electromagnetically technologies which require the use of microscopic or nanoscopic materials, structures, and devices. These all enlist small superconductors in 3D, 2D, 1D, and 0D, which includes bulk like systems, layered or thin film or atomic sheet systems (Osofsky, Hernández-Hangarter, et al., 2016; Osofsky, Krowne, et al., 2016), nanowires/nanotubes/nanocables (Krowne, 2011; Martinez, Calle-Vallejo, Krowne, & Alonso, 2012; Martinez, Abad, Calle-Vallejo, Krowne, & Alonso, 2013), and quantum dots. Obtaining superconductors in such small systems necessitates relaxed requirements on critical values of temperature T_c, current J_c, and magnetic field H_{c1} and H_{c2}. One way to obtain larger values of all these critical constants, may be to introduce disorder intentionally.

We have found that functionalizing graphene (Osofsky, Hernández-Hangarter, et al., 2016) with atoms of F, N, or O, allows one to dial into various matter states, which may be metallic, insulating, or even superconducting. This ability to dial in such characteristics at the nanoscopic or atomic level is something that was not attainable previously in a systematic fashion, which can be repeatedly duplicated now in the laboratory.

This is the latest era of atomic and nanoscopic science: dimensions manipulated are on the order of a few angstroms. One of the remarkable facets of that study is despite the excitement generated by the achievement of metallic single layer graphene, it has been confused by the fact that seminal theoretical work (Abrahams, Anderson, Licciardello, & Ramakrishnan, 1979) predicted that purely two-dimensional (2D) systems should not be metallic − because of disorder. The situation is confounded further by later theoretical work showing that Dirac Fermionic systems with no spin−orbit interactions and Gaussian correlated disorder exhibit scaling behavior, should always be metallic (Das Sarma, Adam, Hwanf, & Rossi, 2011). So what is it, metallic or insulating? Really the answer is both due to the metal−insulator transition (MIT), which also suggests the possibility of dialing into a superconducting state.

Extremely thin metallic-like oxides, on the scale of ten nanometers, such as RuO_2 thin film layers (Osofsky, Krowne, et al., 2016), is a highly disordered conductor in which resistivity does not decrease with decreasing temperature, and the disorder-driven MIT is then described as a quantum phase transition characterized by extended states for the metallic phase and by localized states for the insulating phase. The existence of metallic behavior reported for this 2D material once again violates the famous prediction of Abrahams et al. (1979) that all 2D systems must be localized regardless of the degree of disorder. The discovery of a metallic state in high-mobility metal oxide field-effect transistors (HMFET) motivated several theoretical approaches that included electron−electron interactions to screen disorder. These models adequately described the results for low carrier concentration, high-mobility systems, but are not applicable to the case of highly disordered 2D metals. And once again, the metal−insulator transition (MIT), suggests the possibility of dialing into a superconducting state.

Dimensions for RuO_2 nanoparticles which decorate SiO_2 inner core nanowires, forming nanocables (Krowne, 2011), are on the order of 3 nm. These 1D type of structures, might be investigated for possibilities of superconductivity. Although modeling makes some substantial simplifications, looking at the continuous cylindrical coverage of RuO_2 as a nanotube, in fact it is composed of somewhat randomly placed popcorns of RuO_2 nodules. Although early papers, for example, the one by Anderson (1959), have made some qualitative arguments for disorder effects based upon some first and second order quantum mechanical perturbation theory, the theoretical and experimental insights dependent upon atomic and nanoscopic engineering of materials, did not exist back then, and so limits the reach of such earlier works.

However, ultrasmall samples used in tunneling experiments, led to studies calling into question use of the grand canonical ensemble to describe electron pairing gap $\Delta_0(n_e)$, using an attractive Hubbard model (in real space) to represent s-wave superconductors, obtaining the gap parameter $\Delta_{N_e}(n_e, N)$ varying with averaged electron density n_e and site number N (Tanaka & Marsiglio, 2000a). In Tanaka and Marsiglio (2000a), and in Tanaka and Marsiglio (2000b), the suggested approach in Anderson (1959), solving for the eigenvalues and eigenstates of a non–interacting problem, diagonalizing the single–particle Hamiltonian, finding the transformed electron–electron interaction, and then applying the BCS variational procedure, generates a modified BCS gap equation. An alternative view using the effective Hamiltonian, diagonalizing by a Bogoliubov–Valatin transformation with a de Gennes approach, allows inspection of specific sites with one added impurity atom (Tanaka & Marsiglio, 2000b). Such studies point up the importance of looking at the various nanoscopic aspects of superconductivity, although in the work here, we will not address such site by site nonuniformities (with either BCS and/or BdG tactic).

Superconductivity can be modified by various effects related to randomness, disorder, structural defects, and other similar physical effects. Their affects on superconductivity is important because such effects are intrinsic to certain material system's preparation, or may be intentionally produced. For example, in the 1980s on Anderson localization and dirty superconductors H_{c2} is affected (Kotliar & Kapitulnik, 1986); and high T_c superconductors with structural disorder affects the electron–electron attraction stipulated by exchange of low–energy excitations, with substantial enhancement of T_c (Maleyev & Toperverg, 1988). In the 1990s, looking at thermal fluctuations, quenched disorder, phase transitions, and transport in type-II superconductors, vortices are pinned due to impurities or other defects which destroys long range correlations of the vortex lattice (Fisher, Fisher, & Huse, 1991); in field-induced superconductivity in disordered wire networks, small transverse magnetic applied fields increased the mean T_c in disordered networks (Bonetto, Israeloff, Pokrovskiy, & Bojko, 1998); for structural disorder and its effect on the superconducting transition temperature in the organic superconductor κ-(BEDT-TTF)$_2$Cu[N(CN)$_2$]Br, T_c is reduced in quenched cooled state (Su, Zuo, Schlueter, Kelly, & Williams, 1998); enhancement of J_c density in single-crystal Bi$_2$Sr$_2$CaCu$_2$O$_8$ superconductors occurs by chemically induced disorder (Wang, Wu, Chen, &

Lieber, 1990). In the 2000s, surface enhancement of superconductivity occurred in single crystal tin, due to cold worked surface with surface enhanced order parameter (Kozhevnikov et al., 2005); for disorder and quantum fluctuations in superconducting films in strong magnetic fields, H_{c2} can increase and especially at low temperature (Galitski & Larkin, 2001); preparing amorphous MgB_2/MgO superstructures which produces a model disordered superconductor, bilayers were made with relatively high T_c (Siemons et al., 2008); disorder-induced superconductivity can be produced in ropes of carbon nanotubes, with T_c increasing with disorder (Bellafi, Haddad, & Charfi-Kaddour, 2009); disordered 2D superconductors are examined for the role of temperature and interaction strength in the Hubbard model when the on-site attraction is switched off on a fraction f of sites while keeping the attraction U on the remaining sites, showing that near $f = 0.07$, T_c increases with U in the $2 \leq U \leq 6$ range (Mondaini, Paiva, dos Santos, & Scalettar, 2008); enhancement of the high magnetic field J_c of superconducting MgB_2 by proton irradiation occurs, with the irradiation pinning the vortices increasing J_c (Bugoslavsky et al., 2001); examination of the Lindemann criterion and vortex phase transitions in type-II superconductors, shows the destruction of vortex order by random point pinning and thermal fluctuations (Kierfeld & Vinokur, 2004); doping induced disorder and superconductivity properties in carbohydrate doped MgB_2, increases the J_c density (Kim et al., 2008); the interplay between superconductivity and charge density waves is affected by disorder (Attanasi, 2008); insensitivity of d-wave pairing to disorder in the high temperature cuprate superconductors, increase then decrease from scattering with weak dependence of T_c on n_{imp} impurity in theory as compared to experimental results (Kemper et al., 2009).

In the 2010s, for strongly disordered TiN and NbTiN s-wave superconductors probed by microwave electrodynamics (Driessen, Coumou, Tromp, de Visser, & Klapwijk, 2012), it is mentioned that a decreased T_c with increasing sheet resistance is found for MoGe films by Finkelstein (1987); dynamical conductivity across the disorder tuned superconductor–insulator transition, has disorder enhanced absorption in conductivity and expands the quantum critical region (Swanson, Loh, Randeria, & Trivedi, 2014); effects of randomness on T_c in quasi-2D organic superconductors, leads to lowered T_c (Nakhmedov, Alekperov, & Oppermann, 2012).

Here we show, in the context of a Cooper instability relationship, that introduction of disorder through impurities could possibly lead to an increase in T_c, using one of the simplest reductions from the hierarchy of complex representations. Other reductions in the hierarchy may show increases then decreases, depending on the specific reductions used. Although there is a complicated collection of quantum Green's functions, polarization functions (correlation functions), and vertex functions, used in the hierarchy of assumptions to extract out an analytical formula for T_c, one may obtain at least one formula from it which is recognizable, and relatable to BCS theory. Handling modified BCS superconductors, medium temperature superconductors, and high T_c superconductors may be possible by appropriate use of the more accurate function representations and appropriate adaptions in the hierarchy.

Below in the following sections will be discussed the phonon operator utilized (Section 2), the electron–phonon interaction potential, the Matsubara quantum many-body Green's functions employed for phonons and electrons (Section 3), determination of the perturbed electron Matsubara quantum many-body Green's function from its bare value, an examination of how the perturbed phonon quantum Green's function may be handled in a simplified manner and ramifications (Section 4), finding the renormalized phonon Green's function due to electron screening (Section 5), electron vertex renormalization due to phonons (Section 6), renormalized total potential interaction energy due to Coulomb and phonon effects (Section 7), how the RPA permittivity is controlled through the charge–charge polarization diagram and related issues (Section 8), disorder characterization for impurity scattering (Section 9), obtaining the ladder superconducting Cooper vertex (Section 10) with the incorporation of impurity scattering, and a Cooper instability equation which is solvable in a hierarchy of possible equations of the critical temperature T_c (Section 11). Section 12 relates the gap parameter to the disorder potential energy, giving the indirect relationship between T_c and gap parameter Δ when obtaining Δ's dependence on the same disorder potential energy quantity as T_c. Section 13 offers conclusions.

All of this analysis is done while not necessarily maintaining band structure symmetry at various levels of development. We note that the interested reader can refer to earlier works involving elementary excitations in solids, quasiparticles, and many-body effects in Pines (1964, 1979) and Ginzburg and Kirzhnits (1982), and Kittel (1987) (which includes magnons). Those interested in a more refined electron–phonon interaction model (beyond the two-valued step function model) using the spectral density $\alpha^2(\omega)F(\omega)$, can look at the very thorough review of Corbette (1990), which provides detailed information on the real and imaginary axis analytical continuation relation between, of the solution to the Eliashberg (1960, 1961) equations for the non-weak or strong coupling regimes. It is interesting that with some slight modifications, the strong case for the T_c expression is similar to the weak form, which is treated herein. See Bartolf (2016) for a recent discussion of this and other topics in superconductivity, such as Ginzburg–Landau theory, the two fluid model of Gorter and Casimir, type-II superconductivity with vortices − including work by Abrikosov (2004), Karnaukhov and Shepelev (2008), and the London theory; Tinkham (1980), Schrieffer (1964) for details on the microscopic approach, as well as Abrikosov, Gorkov, and Dzyaloshinski (1963, 1965) and Abrikosov and Gor'kov (1959a) (electrodynamics using a Matsubara thermodynamic approach for δ, relationship between $\mathbf{j}$ and $\mathbf{A}$ [i.e., Q hitting $\mathbf{A}$], and penetration depth δ), Abrikosov and Gor'kov (1959b) (obtains $T = 0$ electrodynamics equations, with introduction of the $\mathbf{A}$ photonic field Green's function D and the system fermions Green's functions G and F, again finding Q and δ), Abrikosov and Gor'kov (1961) (magnetic type impurities are expected to break the time reversal symmetry of the Cooper pairs, so although this is interesting work, our intent in here is to use materials which do not intentionally have such symmetry breaking). For a review of localized impurity states, see Balatsky, Vekhter, and Zhu (2006).

2. PHONON OPERATOR AND THE ELECTRON–PHONON INTERACTION

The complete quantum phonon operator in second quantization format A, requires a raising $b^\dagger$ and lowering operator b. Denoting the reciprocal space phonon momentum as $\mathbf{q}$, with $\mathbf{G}$ being the reciprocal lattice Umklapp vector, the A operator is simply the sum of the two second quantization operators, with appropriate momentum indices.

$$A_{\mathbf{q}+\mathbf{G},\lambda_b,\lambda_p} = b^\dagger_{-(\mathbf{q}+\mathbf{G}),\lambda_b,\lambda_p} + b_{\mathbf{q}+\mathbf{G},\lambda_b,\lambda_p} \tag{1}$$

Here λ_b and λ_p are respectively the phonon branch type (acoustic or optical) and phonon polarization type (longitudinal or transverse [two possibilities]). $b^\dagger$ and b exist in the phonon Hilbert space. In general, the raising and lowering operators for a particular phonon quantum state depend on the surrounding sea of other phonons, so $b^\dagger$ and b differ from their bare operators $b_0^\dagger$ and b_0, which exist without such influences or interactions.

This point will become very important later on in deriving an acceptable form for the overall problem at hand.

Recognizing that the product space $|\psi_{el}\rangle \otimes |\psi_{ph}\rangle$ must be used for the electron–phonon interaction energy $V_{e\text{-}ph}$,

$$
V_{e\text{-}ph} = \frac{1}{\mathcal{V}} \sum_{\mathbf{k} \in FBZ_e} \sum_{\sigma} \sum_{\mathbf{q} \in FBZ_p} \sum_{\lambda_b, \lambda_p} \sum_{\mathbf{G} \in RL} g_{\mathbf{q}, \mathbf{G}, \lambda_b, \lambda_p} c^{\dagger}_{\mathbf{k}+\mathbf{q}+\mathbf{G}, \sigma} c_{\mathbf{k}, \sigma} A_{\mathbf{q}+\mathbf{G}, \lambda_b, \lambda_p} \qquad (2)
$$

where $g_{\mathbf{q}, \mathbf{G}, \lambda_b, \lambda_p}$ is the degeneracy factor for the phonons, $c^{\dagger}$ and c are the destruction and creation electron operators in their electron Hilbert space, and $\mathcal{V}$ is the real space volume. The summations occur respectively for $\mathbf{k}$, σ, $\mathbf{q}$, and $\mathbf{G}$ in the 1st Brillouin zone (FBZ_e) of the electron reciprocal lattice, the two electron spin polarizations, in the 1st Brillouin zone (FBZ_{ph}) for phonons, and in the reciprocal lattice.

3. MATSUBARA MANY-BODY QUANTUM GREEN'S FUNCTIONS FOR PHONONS AND ELECTRONS

The imaginary time based Matsubara quantum many-body Green's functions can be convenient to use, and we do that here for both the phonon and electron propagators. First look at the phonon propagator,

$$
D_{\lambda}(\mathbf{q}, \tau) = -\langle T_{\tau}[A_{\mathbf{q}, \lambda}(\tau) A_{-\mathbf{q}, \lambda}(0)]\rangle_0 \qquad (3)
$$

where the compressed notation form $\lambda = (\lambda_b, \lambda_p)$ is used, and the Umklapp process is reduced to within the 1st Brillouin zone, i.e., $\mathbf{G} = 0$. T_{τ} is the time ordering operator in the imaginary Matsubara time τ frame. Subscript "0" on the outside of the brackets indicates thermal averaging over the ensemble.

Next turn ones attention to the electron propagator, expressing it again in terms of the imaginary time τ,

$$
G_{\sigma}(\mathbf{k}, \tau) = -\frac{\langle T_{\tau}\{U(\beta, 0) c_{\mathbf{k}, \sigma}(\tau) c^{\dagger}_{\mathbf{k}, \sigma}(0)\}\rangle_0}{\langle U(\beta, 0)\rangle_0} \qquad (4)
$$

where $U(\beta, 0)$ is the unitary evolution operator in the imaginary Matsubara time frame, evaluated at $\tau = \beta$. It depends on the time ordering of the perturbing interaction energies $P(\tau_i) = V_{e\text{-}ph}(\tau_i)$. Here $\beta = \hbar/(k_B T)$. The evolution operator for an arbitrary time τ is

$$U(\tau,0) = \sum_{m=0}^{\infty} \frac{1}{m!}(-1)^m \int_0^{\tau} d\tau_1 \int_0^{\tau} d\tau_2 \cdots \int_0^{\tau} d\tau_m T_\tau \{P(\tau_1)P(\tau_2)\cdots P(\tau_m)\}$$

(5)

This formula has assumed the perturbation energy potential acts nearly instantaneously, which is why the $P(\tau_i)$'s are a single function of imaginary time τ_i; $i = 1, 2, 3, \cdots, m$. But, in fact, for phonons as seen by two explicit similarly expressed forms for the pure electron–electron Coulomb interaction $W^{Coul}(\tau)$ and the electron–electron mediated phonon interaction $P(\tau_i, \tau_j) = V_{e\text{-}ph}(\tau_i, \tau_j)$,

$$V_{e\text{-}ph}(\tau_i, \tau_j) = \frac{1}{2\nu} \sum_{\mathbf{k}_1,\sigma_1} \sum_{\mathbf{k}_2,\sigma_2} \sum_{\mathbf{q},\lambda} \frac{1}{\nu} |g_{\mathbf{q},\lambda}|^2 D_\lambda^0(\mathbf{q}, \tau_i - \tau_j) c_{\mathbf{k}_1+\mathbf{q},\sigma_1}^\dagger(\tau_j) c_{\mathbf{k}_2-\mathbf{q},\sigma_2}^\dagger(\tau_i)$$

$$\times c_{\mathbf{k}_2,\sigma_2}(\tau_i) c_{\mathbf{k}_1,\sigma_1}(\tau_j)$$

(6)

$$W^{Coul}(\tau) = \frac{1}{2\nu} \sum_{\mathbf{k}_1,\sigma_1} \sum_{\mathbf{k}_2,\sigma_2} \sum_{\mathbf{q}\neq 0} V_{\mathbf{q}} c_{\mathbf{k}_1+\mathbf{q},\sigma_1}^\dagger(\tau) c_{\mathbf{k}_2-\mathbf{q},\sigma_2}^\dagger(\tau) c_{\mathbf{k}_2,\sigma_2}(\tau) c_{\mathbf{k}_1,\sigma_1}(\tau) \quad (7)$$

the phonon mediated interaction takes a finite time to occur compared to the pure Coulomb interaction. Thus, $\Delta\tau^{lt} = \tau_i^{lt} - \tau_j^{lt} \ll \Delta\tau^{ph} = \tau_i^{ph} - \tau_j^{ph}$. Eq. (7) has merely used the fact that photons mediate the Coulomb interaction, and in a nonrelativistic approximation, it is satisfactorily to set $\Delta\tau^{lt} \approx 0$, or to assign one time to τ_i^{lt} and τ_j^{lt}.

Note that the two particle Coulomb interaction operator $W^{Coul}(\tau)$ containing four c electron operators with Coulomb interaction potential energy $V_{\mathbf{q}}$, has $V_{\mathbf{q}}$ replaced by $|g_{\mathbf{q}\lambda}|^2 D(\mathbf{q}, \tau_i - \tau_j)$ for phonon mediated e–e interaction. The phonon mediated e–e interaction has used the bare (unperturbed) phonon Green's function $D_\lambda^0(\mathbf{q}, \tau_i - \tau_j)$, whose justification will be covered next.

4. APPROXIMATION OF THE PHONON MATSUBARA PROPAGATOR BY ITS BARE VALUE

It is worthwhile to make an argument as to why it might be propitious to simplify the Green's function for the phonon from one dependent on the perturbed many-body form to one given by its bare unperturbed form. The guiding reasoning is this. If the complete and general Green's functions for both the electrons and phonons are used, phonon interactions will be renormalized by the many-body electron effects, but also the electron effects will be renormalized by the phonon many-body effects. This would constitute what we could refer to as simultaneous double renormalizations. Clearly, this would be the most accurate way to do the calculations. But we suspect, because the phonons originate from ionic motion, and ions are extremely massive compared to the electron mass, that the electron effect may dominate because of their much higher velocity. Of course, a somewhat counter argument is that it is really the momentum that is important, and that the $p_e = m_e v_e$ product may not exceed that of the phonon momentum p_{ph}, because $M_{ion} \gg m_e$.

A way out of this conundrum is to evaluate the effect of additional phonon lines on an electron ⟶ scatter off of a vertex 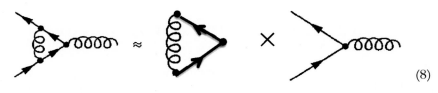 with a phonon , produced namely by

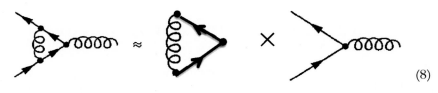

$$(8)$$

where the first product factor on the right hand side is about

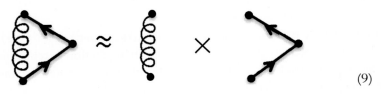

$$(9)$$

Close top and bottom terminations on the right most product on the right hand side of (9) as an approximation, giving us a pair bubble times the bare vertex:

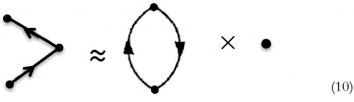

$$(10)$$

Here the polarization bubble (Fetter & Walecka, 1971) is

$$= \chi_0(q, iq_n) = -\Pi_0(q, iq_n) = \chi_0^R(q, 0)\Big|_{\substack{low\ T \\ long\ \lambda_q}} = \chi_0(q, 0) \approx -d(\varepsilon_F) \tag{11}$$

Here $d(\varepsilon_F)$ is the density of states at the Fermi energy (units are in inverse *energy · volume*). The bare phonon propagator D_λ^0, with its effective vertex strength C_ν^{eff}, is

$$= D_\lambda^0(\mathbf{q}, iq_n) = \frac{2\hbar\omega_q}{(iq_n)^2 - (\hbar\omega_q)^2}\bigg|_{|iq_n| \ll \omega_q < \omega_D} \approx -\frac{2}{\hbar\omega_q} \tag{12}$$

and

$$C_\nu^{eff} = \frac{1}{\mathcal{V}}|g_q|^2 \times \frac{\hbar\omega_D}{\varepsilon_F} = \frac{1}{2}W^{Coul}(q)\hbar\omega_q \times \frac{\hbar\omega_D}{\varepsilon_F}$$
$$\approx \frac{1}{2}W^{R,Coul}(q)\Big|_{\substack{q \to 0, \\ iq_n=0}} \hbar\omega_q \times \frac{\hbar\omega_D}{\varepsilon_F} \approx \frac{1}{2d(\varepsilon_F)}\hbar\omega_q \times \frac{\hbar\omega_D}{\varepsilon_F} \tag{13}$$

Result for the correction factor is (using Bohm–Staver expression for sound v_s)

$$\approx \frac{\hbar\omega_D}{\varepsilon_F} = \frac{\hbar\omega_D}{p_F v_F/2} = \frac{\hbar\omega_D}{\hbar k_F v_F/2} = \frac{v_s k_D}{k_F v_F/2} = \sqrt{\frac{Z m_e}{3 M_{atom}}} \frac{k_D}{k_F v_F/2}$$

$$= 2\sqrt{\frac{Z}{3}}\sqrt{\frac{m_e}{M_{atom}}}\frac{k_D}{k_F} \approx \sqrt{\frac{m_e}{M_{atom}}} \tag{14}$$

giving for the correction term,

$$\approx \sqrt{\frac{m}{M}} \times \tag{15}$$

and showing, because $m_e/M_{ion} \ll 1$, that the renormalization process, may be considerably simplified, by dropping terms with vertices dressed by additional phonon lines. Such dropping of those terms is known as Migdal's theorem, and we enlist it to approximate the fully dressed (perturbed) phonon propagator D_λ, with the bare (unperturbed) propagator D_λ^0.

5. FINDING THE RENORMALIZED PHONON GREEN'S FUNCTION DUE TO ELECTRON SCREENING

The phonon propagator may be renormalized in the RPA approximation, denoted by $D_\lambda^{RPA,ren}(\mathbf{q}, iq_n)$ or by $D_\lambda^{RPA}(\mathbf{q}, iq_n)$ for short. Diagrammatically, it is expressible in a Dyson equation form, that is it is equal to the bare phonon value plus a correction due to the RPA effect of the electron screening

$$- D_\lambda^{RPA}(\mathbf{q}, iq_n)$$

$$= \mathord{\text{〰〰〰}} = \mathord{\text{〰〰〰}} + \mathord{\text{〰〰〰}} \!\! \boxed{\text{RPA}} \!\! \mathord{\text{〰〰〰}} \tag{16}$$

This Dyson like equation may be solved for the renormalized phonon Green's function:

$$-D_\lambda^{RPA}(\mathbf{q}, iq_n) = \mathord{\text{〰〰〰〰}} = \frac{\mathord{\text{〰〰〰}}}{1 - \boxed{\text{RPA}}\,\mathord{\text{〰〰〰}}} \tag{17}$$

Phonon vertices ●, from left to right, attached to the $-\chi^{RPA}(\tilde{q})$ electron screening RPA bubble ●$\boxed{\text{RPA}}$●, are given by respectively, $g_{\mathbf{q}\lambda}/\sqrt{\mathcal{V}}$ and $g_{\mathbf{q}\lambda}^*/\sqrt{\mathcal{V}}$. Eq. (17) can be rewritten as

$$D_\lambda^{RPA}(\mathbf{q}, iq_n) = \frac{D_\lambda^0(\mathbf{q}, iq_n)}{1 - \frac{1}{\mathcal{V}}|g_{\mathbf{q}\lambda}|^2 D_\lambda^0(\mathbf{q}, iq_n)\chi^{RPA}(\mathbf{q}, iq_n)} \tag{18}$$

This expression may be evaluated with the bare phonon Green's function $D_\lambda^0(\mathbf{q}, iq_n)$,

$$D_\lambda^0(\mathbf{q}, iq_n) = \frac{2\Omega_{\mathbf{q}\lambda}}{(iq_n)^2 - (\Omega_{\mathbf{q}\lambda})^2} \tag{19}$$

and the vertex products under the Jellium model,

$$\frac{1}{\mathcal{V}}|g_{\mathbf{q}\lambda}|^2 = \left(\frac{1}{\sqrt{\mathcal{V}}}g_{\mathbf{q}\lambda}\right)\left(\frac{1}{\sqrt{\mathcal{V}}}g_{\mathbf{q}\lambda}\right) = \frac{1}{2}W^{Coul}(q)\Omega_{\mathbf{q}\lambda} \tag{20}$$

Renormalized phonon propagator is then,

$$D_\lambda^{RPA,ren}(\mathbf{q}, iq_n) = \frac{2\Omega_{\mathbf{q}\lambda}}{(iq_n)^2 - (\Omega_{\mathbf{q}\lambda})^2[1 + W^{Coul}(q)\chi^{RPA}(\mathbf{q}, iq_n)]} \qquad (21)$$

Square bracketed expression is the inverse of the 4-reciprocal space RPA renormalized permittivity,

$$\varepsilon^{RPA}(\mathbf{q}, iq_n) = \frac{1}{1 + W^{Coul}(q)\chi^{RPA}(\mathbf{q}, iq_n)} \qquad (22)$$

Final expression for renormalized phonon propagator is then,

$$D_\lambda^{RPA,ren}(\mathbf{q}, iq_n) = \frac{2\Omega_{\mathbf{q}\lambda}}{(iq_n)^2 - (\Omega_{\mathbf{q}\lambda})^2/\varepsilon^{RPA}(\mathbf{q}, iq_n)} = \frac{2\Omega_{\mathbf{q}\lambda}}{(iq_n)^2 - (\omega_{\mathbf{q}\lambda}^{ren})^2} \qquad (23)$$

with the renormalized phonon frequency given by

$$\omega_{\mathbf{q}\lambda}^{ren} = \frac{\Omega_{\mathbf{q}\lambda}}{\sqrt{\varepsilon^{RPA}(\mathbf{q}, iq_n)}} \qquad (24)$$

In the Jellium model, $\Omega_{\mathbf{q}\lambda}$ is actually independent of both $\mathbf{q}$ and λ, given by

$$\Omega_{\mathbf{q}\lambda}|_{Jellium} = \Omega = \sqrt{\frac{Z^2 e^2 N_{ion}}{\varepsilon_0 M_{ion}\mathcal{V}}} = \sqrt{\frac{Z^2 e^2 (\rho_{ion}^0 \mathcal{V})}{\varepsilon_0 M_{ion}\mathcal{V}}} = \sqrt{\frac{Z^2 e^2 \rho_{ion}^0}{\varepsilon_0 M_{ion}}}$$

$$= \sqrt{\frac{Ze^2 (Z\rho_{ion}^0)}{\varepsilon_0 M_{ion}}} = \sqrt{\frac{Ze^2 \rho_{el}^0}{\varepsilon_0 M_{ion}}} \qquad (25)$$

6. RENORMALIZED ELECTRON VERTEX BASED ON PHONON MODIFICATION

We denote the modified bare vertex ●, by , and see that it is expressed as

$$\text{} = \frac{1}{\sqrt{\mathcal{V}}}g_{\mathbf{q}\lambda}^{RPA,ren} = \left[1 + \bullet \text{〰〰〰} \left(\boxed{RPA}\right) \bullet \right]\frac{1}{\sqrt{\mathcal{V}}}g_{\mathbf{q}\lambda} \qquad (26)$$

Inserting the expression for the phonon line and the RPA bubble into (26),

$$\bigotimes = \frac{1}{\sqrt{\mathcal{V}}}g_{\mathbf{q}\lambda}^{RPA,ren} = \left[1 + \{-W(q)\}\{-\chi^{RPA}(\tilde{q})\}\right]\frac{1}{\sqrt{\mathcal{V}}}g_{\mathbf{q}\lambda} \qquad (27)$$

And stripping away the extra $1/\sqrt{\mathcal{V}}$ factors, we finally find the phonon mediated electron–electron vertex,

$$\begin{aligned} g_{\mathbf{q}\lambda}^{RPA,ren} &= \left[1 + W(q)\chi^{RPA}(\tilde{q})\right]g_{\mathbf{q}\lambda} \\ &= \frac{g_{\mathbf{q}\lambda}}{\varepsilon^{RPA}(\mathbf{q}, iq_n)} \end{aligned} \qquad (28)$$

which shows a renormalization by the permittivity.

7. RENORMALIZED TOTAL POTENTIAL ENERGY DUE TO COULOMB AND PHONON EFFECTS

There exists an effective potential energy $V_{eff}^{RPA,ren}$ which combines the effect of the repulsive Coulomb energy, RPA renormalized, and the attractive electron–electron phonon mediated energy, RPA renormalized,

$$V_{eff}^{RPA,ren} = W^{Coul,RPA,ren}(\mathbf{q}, iq_n) + \frac{1}{\mathcal{V}}\left|g_{\mathbf{q}\lambda}^{RPA,ren}\right|^2 D_\lambda^{RPA,ren}(\mathbf{q}, iq_n) \qquad (29)$$

Diagrammatically, it looks like

$$\text{⟞⟝⟞⟝⟞⟝} = \text{⟞⟝⟞⟝} + \text{⟞⟝⟞⟝⟞⟝} \qquad (30)$$

where the renormalized phonon propagator with its vertices is

$$\begin{aligned} \text{⟞⟝⟞⟝} &= -\frac{1}{\mathcal{V}}|g_{\mathbf{q}\lambda}|^2 D_\lambda^{RPA}(\mathbf{q}, iq_n) = -\frac{W^{Coul}(q)}{\varepsilon^{RPA}(\tilde{q})}\frac{(\omega_{\mathbf{q}\lambda}^{ren})^2}{(iq_n)^2 - (\omega_{\mathbf{q}\lambda}^{ren})^2} \\ &= -W^{Coul,RPA,ren}(\tilde{q})\frac{(\omega_{\mathbf{q}\lambda}^{ren})^2}{(iq_n)^2 - (\omega_{\mathbf{q}\lambda}^{ren})^2} \end{aligned} \qquad (31)$$

and where the renormalized Coulomb interaction is identified as

$$W^{Coul,RPA,ren}(\tilde{q}) = \frac{W^{Coul}(q)}{\varepsilon^{RPA}(\tilde{q})} \qquad (32)$$

Reduction of the bare Coulomb effect seen in (32) is like that in earlier works (Ginzburg & Kirzhnits, 1982), (1.28) and (4.1) giving, respectively, $U_c^* = U_c[1 + U_c \ln(\omega_F/\omega_c)]^{-1}$, $\mu^* = \mu[1 + U_c \ln(E_F/\bar{\omega})]^{-1}$; change $\bar{\omega}$ to ω in Kittel (1987) for (2.36).

8. RPA PERMITTIVITY AS AFFECTED BY THE CORRELATION FUNCTION POLARIZATION

RPA permittivity is controlled through the correlation function, the charge–charge polarization χ. Its expression in reciprocal 4-space is (Fetter & Walecka, 1971)

$$\varepsilon^{RPA}(\mathbf{q}, iq_n) = 1 - \frac{e^2}{\varepsilon_0 q^2} \chi_0(\mathbf{q}, iq_n) \tag{33}$$

What happens at the low frequency limit? By analytic continuation, the retarded polarization χ^R at any frequency ω is

$$\chi_0^R(\mathbf{q}, \omega) = \chi_0(\mathbf{q}, iq_n \to \omega + i\eta); \quad \eta = 0^+ \tag{34}$$

However, it is known that as $\omega \to 0$ that

$$\chi_0^R(\mathbf{q}, \omega)\big|_{\omega \to 0} \approx -d(\varepsilon_F) \tag{35}$$

where d is the density of states (DOS) at the Fermi level ε_F, given in $energy^{-1} \cdot volume^{-1}$ units. But in the same limit,

$$\chi_0^R(\mathbf{q}, \omega)\big|_{\omega \to 0} = \chi_0(\mathbf{q}, 0) \tag{36}$$

which makes

$$\chi_0(\mathbf{q}, 0) \approx -d(\varepsilon_F) \tag{37}$$

So the polarization we need in (33) is determined by the density of states at the Fermi level ε_F, which in 3D systems looks like

$$d^{3D}(\varepsilon_F) = \frac{1}{2\pi^2}\left(\frac{2m_e}{\hbar^2}\right)^{3/2}(\varepsilon_F)^{1/2} = \frac{m_e k_F}{\pi^2 \hbar^2}; \quad \varepsilon_F = \frac{\hbar^2 (k_F)^2}{2m_e} \tag{38}$$

making the Fermi–Thomas screening wavenumber, squared,

$$\left(k_s^{3D}\right)^2 = -4\pi (e_0)^2 \chi_0^{3D}(0, 0) = \frac{4}{\pi}\frac{k_F}{a_0}; \quad e_0 = \frac{\hbar^2}{m_e a_0} \tag{39}$$

RPA permittivity is then

$$\varepsilon^{RPA,3D}(\mathbf{q}, iq_n)\big|_{iq_n \to 0} \approx 1 + \frac{(k_s^{3D})^2}{q^2} = 1 + \frac{4k_F}{\pi a_0} \frac{1}{q^2} \tag{40}$$

In the small wavelength limit $q = 2\pi/\lambda$, q becomes very large, and $\varepsilon^{RPA,3D}(\mathbf{q}, iq_n)\big|_{iq_n \to 0} \to 1$, or permittivity goes to unity. In the opposite limit for large wavelengths, q approaches zero, and

$$\varepsilon^{RPA,3D}(\mathbf{q}, iq_n)\big|_{\substack{\mathbf{q} \to 0 \\ iq_n \to 0}} \approx \frac{(k_s^{3D})^2}{q^2} = \frac{4k_F}{\pi a_0} \frac{1}{q^2} \tag{41}$$

This makes renormalized phonon frequency

$$\omega_{\mathbf{q}\lambda}^{ren} = \sqrt{\frac{Ze^2 \rho_{el}^0}{\varepsilon^{RPA}(\mathbf{q}, iq_n)\varepsilon_0 M_{ion}}} \tag{42}$$

become, using the formula for electron density

$$\rho_{el}^0 = n^{3D} = \frac{(k_F)^3}{3\pi^2} \tag{43}$$

the low frequency large wavelength result, with a linear frequency vs. momentum q relationship,

$$\omega_{\mathbf{q}\lambda}^{ren} = \sqrt{\frac{Zm_e}{3M_{ion}}} v_F q = v_{s,ac}q; \quad v_{s,ac} = \sqrt{\frac{Zm_e}{3M_{ion}}} v_F \tag{44}$$

This gives us the Bohm–Staver relationship, for the acoustic dispersion of the phonons, from the Jellium model. The original optical Jellium phonons, get screened, and reduced to low frequencies. This happens because the RPA permittivity is enormous, proportional to $1/q^2$, and approaches infinity as q goes to 0.

For 2D ordinary metals, and atomic layered graphene, various changes have to occur, including in χ, to obtain the various quantities like $\varepsilon^{RPA,2D}(\mathbf{q}, iq_n)$ and $\omega_{\mathbf{q}\lambda}^{ren,2D}$. For 2D ordinary metals, the DOS is a constant,

$$\rho^{2D}(\varepsilon) = \frac{g_{s,b}m^*}{2\pi\hbar^2} \tag{45}$$

where $g_{s,b}$ is the degeneracy due to spin and bands, either conduction or valence bands. In contrast to ordinary 2D metals, graphene with a single atomic layer of carbon atoms, has an energy dependent DOS like 3D systems. However, the power dependence on energy converts from a square root to linear form, in a tight-binding model approach,

$$\rho^{graphene}(\varepsilon) = \frac{g_{s,c}|\varepsilon|}{2\pi\hbar^2 v_F^{gr}} \tag{46}$$

with $\varepsilon > 0$ for the top Dirac cone and $\varepsilon < 0$ for the bottom Dirac cone, and its Fermi velocity given by an overlap orbital nearest neighbor construction,

$$v_F^{gr} = -\frac{3t_{gr}^{nn}a_{C-C}}{2\hbar} = \frac{3|t_{gr}^{nn}|a_{C-C}}{2\hbar} = 9.7 \times 10^7 \text{ cm/s} \tag{47}$$

Here $g_{s,c}$ is the degeneracy due to spin and cones (6 for either the top or bottom Dirac cones), a_{C-C} the nearest neighbor carbon-to-carbon atom spacing, and the energy

$$\bar{\varepsilon}_{\mathbf{q}}^{\lambda_c} = \lambda_c \hbar v_F q; \quad q = |\mathbf{q}|; \quad \lambda_c = \pm 1 \tag{48}$$

where λ_c is the cone type, $+1$ for upper, -1 for lower. This is the energy with respect to the single unique onsite energy of the C atoms, ε_{onsite}, the quantity utilized in (46).

9. DISORDER CHARACTERIZED IMPURITY SCATTERING

Disorder can be described by scattering from charged impurity ions, but nonmagnetic, or spinless. The bare electron–ion interaction is represented by a single dashed line going between a vertex and a star symbolizing the ion (say the jth one here). The renormalized interaction is then expressed as

$$u_j^{RPA}(\mathbf{q}) = \bullet\!=\!=\!=\!\bigstar\!=\!\bullet\!-\!-\!-\!\bigstar + \bullet\!\sim\!\!\!\bigcirc\!\!\!=\!\bigstar \tag{49}$$

and written out as an equation,

$$u_j^{RPA}(\mathbf{q}) = u_j(\mathbf{q}) + \{-W^{Coul}(\mathbf{q})\}\{-\chi_0(\mathbf{q})\}u_j^{RPA}(\mathbf{q}) \tag{50}$$

This Dyson form of equation, can be readily solved, yielding

$$u_j^{RPA}(\mathbf{q}) = \frac{u_j(\mathbf{q})}{1 - W^{Coul}(\mathbf{q})\chi_0(\mathbf{q})}$$

$$= \frac{u_j(\mathbf{q})}{\varepsilon_0^{RPA}(\mathbf{q})} \tag{51}$$

using a single bubble in the RPA expansion for χ. Here the potential at the jth ion is found by lattice displacements $\mathbf{P}_j$,

$$u_j(\mathbf{q}) = u(\mathbf{q})e^{-i\mathbf{q}\cdot\mathbf{P}_j} \tag{52}$$

It is helpful to clarify at this point precisely what is meant by the RPA or random phase approximation. What it is a solution for the general sum of all diagrams for the correlation function,

$$-\chi(\tilde{q}) = \qquad = \qquad \tag{53}$$

but with the irreducible diagrams correlation function $\chi^{irr}(\tilde{q})$ replaced in this Dyson solution of $\chi(\tilde{q})$, with a simple pair-bubble, or

$$-\chi^{irr}(\mathbf{q}, iq_n) = \qquad \tag{54}$$

replaced by

$$-\chi_{RPA}^{irr}(\mathbf{q}, iq_n) = \qquad = -\chi_0(\mathbf{q}, iq_n) \tag{55}$$

Solution of (52) depends not only on the simple polarization diagram, but also on $u_j(\mathbf{q})$. For similar impurities all $u_j(\mathbf{q})$ would be the same, perhaps looking like in real space

$$u(\mathbf{r}) = -\frac{e_0^2}{|\mathbf{r}|}e^{-|\mathbf{r}|/a_d} \tag{56}$$

where a_d is the characteristic decay distance. Its Fourier transform $u(\mathbf{q})$ is then

$$u(\mathbf{q}) = \iiint d^3r\, u(\mathbf{r})e^{-i\mathbf{q}\cdot\mathbf{r}} = -\frac{1}{4\pi\varepsilon_0}\frac{e^2}{(1/a_d)^2 + q^2} \tag{57}$$

10. LADDER SUPERCONDUCTING COOPER VERTEX WITH DISORDER INCORPORATED

Superconducting vertex Λ is a ladder sum of the total effective interaction ●〜〜〜●, which includes the renormalized electron–electron, mediated phonon, and impurity scattering effects,

$$ \tag{58} $$

where $V_{eff,tot}^{RPA}$ is the total interaction due to these effects. Eq. (58) can be restated as a Dyson type equation

$$ \tag{59} $$

This vertex Dyson type equation, may be written out explicitly in terms of its vertex functions, quantum Green's functions, and effective total potential energy $V_{eff,tot}^{RPA}$:

$$ \Lambda(\tilde{k}, \tilde{p}) = - V_{eff\ tot}^{RPA}(\tilde{k} - \tilde{p}) + \frac{1}{\nu\beta} \sum_{\tilde{q}, \lambda_{ph}} [-V_{eff\ tot}^{RPA}(\tilde{k} - \tilde{q})] G_{\uparrow}^{0}(\tilde{q}) G_{\downarrow}^{0}(-\tilde{q}) \Lambda(\tilde{q}, \tilde{p}) \tag{60} $$

Noting that $V_{eff,tot}^{RPA} < 0$, writing it as $V' = - V_{eff,tot}^{RPA}$, and dropping the super- and subscripts for brevity,

$$ \Lambda(\tilde{k}, \tilde{p}) = V'(\tilde{k} - \tilde{p}) + \frac{1}{\nu\beta} \sum_{\tilde{q}, \lambda_{ph}} [V'(\tilde{k} - \tilde{q})] G_{\uparrow}^{0}(\tilde{q}) G_{\downarrow}^{0}(-\tilde{q}) \Lambda(\tilde{q}, \tilde{p}) \tag{61} $$

Formally, this may be solved as follows:

$$\Lambda(\tilde{k}, \tilde{p}) = \frac{V'(\tilde{k} - \tilde{p})}{\left\{ 1 - \frac{1}{V\beta} \sum_{\tilde{q}, \lambda_{ph}} [V'(\tilde{k} - \tilde{q})] G_{\uparrow}^{0}(\tilde{q}) G_{\downarrow}^{0}(-\tilde{q}) \frac{\Lambda(\tilde{q}, \tilde{p})}{\Lambda(\tilde{k}, \tilde{p})} \right\}} \tag{62}$$

Eq. (62) demonstrates the instability behavior when the 2nd denominator term approaches unity from the lower or upper sides.

11. CRITICAL TEMPERATURE OBTAINED FROM A COOPER INSTABILITY EQUATION

Let us examine the Cooper instability equation, obtained from Eq. (62) by setting the denominator to unity,

$$1 - \frac{1}{V\beta} \sum_{\tilde{q}, \lambda_{ph}} [V'(\tilde{k} - \tilde{q})] G_{\uparrow}^{0}(\tilde{q}) G_{\downarrow}^{0}(-\tilde{q}) \frac{\Lambda(\tilde{q}, \tilde{p})}{\Lambda(\tilde{k}, \tilde{p})} = 0 \tag{63}$$

which should be evaluated at $T = T_c$, the critical temperature, since we expect the major physical change to occur here, and should be associated with the singularity. For the vertex function depending approximately only on its second variable, here $\tilde{p}$, the instability condition becomes

$$1 - \frac{1}{V\beta_c} \sum_{\tilde{q}, \lambda_{ph}} [V'(\tilde{k} - \tilde{q})] G_{\uparrow}^{0}(\tilde{q}) G_{\downarrow}^{0}(-\tilde{q}) = 0 \tag{64}$$

which may be rewritten as

$$1 = \frac{1}{V\beta_c} \sum_{\substack{iq_n \\ |iq_n| < \alpha\omega_D}} \sum_{q} \sum_{\lambda_{ph}} [V'(\tilde{k} - \tilde{q})] G_{\uparrow}^{0}(\tilde{q}) G_{\downarrow}^{0}(-\tilde{q}) \tag{65}$$

Here q_n is the Fermion index used in Matsubara reciprocal space imaginary frequency summations,

$$q_n = \frac{(2n+1)\pi}{\beta_c} = \frac{2\pi}{\beta_c}\left(n + \frac{1}{2}\right); \quad \beta_c = \frac{1}{k_B T_c} \tag{66}$$

Bare Green's functions are

$$G_{\uparrow}^{0}(\mathbf{q}, iq_n) = \frac{1}{iq_n - \xi_{\mathbf{q}\uparrow}}; \quad G_{\downarrow}^{0}(\mathbf{q}, iq_n) = \frac{1}{iq_n - \xi_{\mathbf{q}\downarrow}} \tag{67}$$

Considering the case with no static or otherwise background magnetic field,

$$\xi_{\mathbf{q}\uparrow} = \xi_{\mathbf{q}\downarrow} = \xi_{\mathbf{q}} \tag{68}$$

Placing (67) and (68) into (65) then gives

$$1 = \frac{1}{\mathcal{V}\beta_c} \sum_{\substack{iq_n \\ |iq_n| < \alpha\omega_D}} \sum_{\mathbf{q}} \sum_{\lambda_{ph}} [V'(\tilde{k} - \tilde{q})] \frac{1}{iq_n - \xi_{\mathbf{q}}} \frac{1}{-iq_n - \xi_{\mathbf{q}}} \tag{69}$$

The disorder effect contributes an extra diagram ●=⇒❋=⇒● for renormalized impurity scattering to the effective potential energy due to the Coulomb scattering offsetting the attractive electron–electron phonon mediated scattering:

$$\tag{70}$$

or

$$V_{eff}^{RPA,ren} = W^{Coul,RPA,ren}(\mathbf{q}, iq_n) + \frac{1}{\mathcal{V}} |g_{\mathbf{q}\lambda}^{RPA,ren}|^2 D_{\lambda}^{RPA,ren}(\mathbf{q}, iq_n)$$
$$+ W^{imp\ 1BA,RPA}(\mathbf{q}, iq_n) \tag{71}$$

where the last term is given in the 1st Born approximation when only scattering off of a single impurity scatter is allowed (Born approximation), and it is limited to one scattering event (1st). Under these assumptions, valid when scattering due to impurities is not too severe, the last term for impurity potential interaction energy is (Bruus & Flensberg, 2004)

$$W^{imp\ 1BA,RPA}(\mathbf{q}, iq_n) = n_{imp} \frac{u(\mathbf{q})}{\varepsilon^{RPA}(\mathbf{q}, 0)} \frac{u(-\mathbf{q})}{\varepsilon^{RPA}(-\mathbf{q}, 0)} \delta_{q_n,0} \tag{72}$$

which we see has the renormalized impurity interaction lines for incoming $\mathbf{q}$ and outgoing $-\mathbf{q}$ momentum into the impurity center vertex, weighted by the impurity spatial density n_{imp}, which arises from the summation over all scatters per unit volume. The Kronecker delta function, forces one to consider Matsubara frequencies approaching zero.

In a hierarchy of approximations, with various physical effect ramifications, one simplification is to pull $V'(\tilde{k} - \tilde{q})$ out from under the $\mathbf{q}$ and

λ_{ph} summation signs, replacing it by an average. This process is consistent with the earlier two variable to one variable vertex function simplification, which suggests here we drop dependence on **k**.

$$1 = \frac{V'_{av}}{\nu\beta_c} \sum_{\substack{iq_n \\ |iq_n|<\alpha\omega_D}} \sum_{\mathbf{q}} \frac{1}{iq_n - \xi_{\mathbf{q}}} \frac{1}{-iq_n - \xi_{\mathbf{q}}} \tag{73}$$

Here the λ_{ph} summation has been dropped entirely (and absorbed into V'_{av}) since the remaining bare Green's function contains no phonon polarization information. The **q** summation is implicitly over one spin type, where $\xi_{\mathbf{q}}$ is taken above ($\xi_{\mathbf{q}} > 0$) and below ($\xi_{\mathbf{q}} < 0$) the Fermi surface. Above the Fermi surface $\xi_{\mathbf{q}}$ is unbounded, but below it can go to, at least in a normal 3D metal, to $-\varepsilon_F$. Therefore, using the density of states $d_{st}(\xi)$, (73) becomes

$$
\begin{aligned}
1 &= \frac{V'_{av}}{\nu\beta_c} \sum_{\substack{iq_n \\ |iq_n|<\alpha\omega_D}} \sum_{\mathbf{q}} \frac{1}{iq_n - \xi_{\mathbf{q}}} \frac{1}{-iq_n - \xi_{\mathbf{q}}} \\
&= \frac{V'_{av}}{\nu\beta_c} \sum_{\substack{iq_n \\ |iq_n|<\alpha\omega_D}} \frac{1}{\nu} \sum_{\mathbf{q}} \frac{1}{iq_n - \xi_{\mathbf{q}}} \frac{1}{-iq_n - \xi_{\mathbf{q}}} \\
&= \frac{V'_{av}}{\beta_c} \sum_{\substack{iq_n \\ |iq_n|<\alpha\omega_D}} \frac{1}{2} \int_{-\varepsilon_F}^{\infty} d_{st}(\xi)d\xi \frac{1}{iq_n - \xi_{\mathbf{q}}} \frac{1}{-iq_n - \xi_{\mathbf{q}}} \\
&= \frac{V'_{av}}{\beta_c} \sum_{\substack{iq_n \\ |iq_n|<\alpha\omega_D}} \frac{1}{2} \int_{-\varepsilon_F}^{\infty} d_{st}(\xi)d\xi \frac{1}{(q_n)^2 + \xi^2} \\
&\approx \frac{V'_{av}}{\beta_c} \sum_{\substack{iq_n \\ |iq_n|<\alpha\omega_D}} d_{st}(\xi=0; \varepsilon=\varepsilon_F) \frac{1}{2} \int_{-\varepsilon_F}^{\infty} d\xi \frac{1}{(q_n)^2 + \xi^2} \\
&\approx \frac{V'_{av}}{\beta_c} \sum_{\substack{iq_n \\ |iq_n|<\alpha\omega_D}} d_{st}(\xi=0; \varepsilon=\varepsilon_F) \frac{1}{2} \int_{-\infty}^{\infty} d\xi \frac{1}{(q_n)^2 + \xi^2} \\
&= \frac{V'_{av}}{\beta_c} \sum_{\substack{iq_n \\ |iq_n|<\alpha\omega_D}} \frac{d_{st}(\varepsilon_F)}{2} \frac{1}{q_n} \tan^{-1}\left(\frac{\xi}{q_n}\right)\Big|_{-\infty}^{+\infty} \\
&= \frac{V'_{av}}{\beta_c} \sum_{\substack{iq_n \\ |iq_n|<\alpha\omega_D}} \frac{d_{st}(\varepsilon_F)}{2} \frac{2}{q_n} \tan^{-1}(\infty)\mathrm{sgn}(q_n)
\end{aligned}
$$

$$= \frac{V'_{av}}{\beta_c} \sum_{\substack{iq_n \\ |iq_n|<\alpha\omega_D}} \frac{d_{st}(\varepsilon_F)}{2} \frac{2}{|q_n|} \frac{\pi}{2} \tag{74}$$

The final result, obtained from the last line of Eq. (74), is a double-sided sum over Matsubara frequencies. The idea is to convert this into a single sided sum, convert that into an integral, and recognize a familiar form which has an analytical reduction. From (74) we write

$$1 = \frac{V'_{av}}{2\beta_c} \sum_{\substack{iq_n \\ |iq_n|<\alpha\omega_D}} d_{st}(\varepsilon_F) \frac{\pi}{|q_n|} = \frac{V'_{av} d_{st}(\varepsilon_F)}{2\beta_c} \sum_{\substack{iq_n \\ |iq_n|<\alpha\omega_D}} \frac{\pi}{|q_n|} \tag{75}$$

[Note that the earlier DOS $d(\varepsilon_F)$, and $d_{st}(\varepsilon_F)$, are identical, with the additional subscript merely serving to emphasize the distinction from the differential operator inside the integral.] Carrying out the steps outlined, using the Fermion frequency index relation (66),

$$1 = \frac{V'_{av} d_{st}(\varepsilon_F)}{2\beta_c} \sum_{\substack{iq_n \\ |iq_n|<\alpha\omega_D}} \frac{\pi}{|q_n|}$$

$$= \frac{V'_{av} d_{st}(\varepsilon_F)}{2\beta_c} \sum_{\substack{iq_n \\ |iq_n|<\alpha\omega_D}} \frac{\pi}{\left|\frac{(2n+1)\pi}{\beta_c}\right|}$$

$$= \frac{V'_{av} d_{st}(\varepsilon_F)}{2\beta_c} \sum_{\substack{iq_n \\ |iq_n|<\alpha\omega_D}} \frac{\beta_c}{2} \frac{1}{|n+1/2|}$$

$$= \frac{V'_{av} d_{st}(\varepsilon_F)}{2\beta_c} \frac{\beta_c}{2} \sum_{\substack{iq_n \\ |iq_n|<\alpha\omega_D}} \frac{1}{|n+1/2|}$$

$$= \frac{V'_{av} d_{st}(\varepsilon_F)}{4} \sum_{\substack{n=0,\pm1,\pm2,\cdots \\ |q_n|<\alpha\omega_D}} \frac{1}{|n+1/2|}$$

$$= \frac{V'_{av} d_{st}(\varepsilon_F)}{4} \left\{ \sum_{\substack{n=0,1,2,\cdots \\ |q_n|<\alpha\omega_D}} \frac{1}{n+1/2} + \sum_{\substack{n=-1,-2,\cdots \\ |q_n|<\alpha\omega_D}} \frac{1}{|n+1/2|} \right\}$$

$$= \frac{V'_{av} d_{st}(\varepsilon_F)}{4} \left\{ \sum_{n=0}^{\alpha\beta_c\omega_D/(2\pi)-1/2} \frac{1}{n+1/2} + \sum_{n=-1}^{-\alpha\beta_c\omega_D/(2\pi)-1/2} \frac{1}{|n+1/2|} \right\}$$

$$= \frac{V'_{av} d_{st}(\varepsilon_F)}{4} 2 \sum_{n=0}^{\alpha \beta_c \omega_D/(2\pi)-1/2} \frac{1}{n+1/2} \tag{76}$$

So we have obtained a single-sided sum as desired,

$$1 = \frac{V'_{av} d_{st}(\varepsilon_F)}{2} \sum_{n=0}^{\alpha \beta_c \omega_D/(2\pi)-1/2} \frac{1}{n+1/2} \tag{77}$$

One way to handle this summation is to replace it by an integral approximation, taking $1 \to dn$, and perform a change of variable, so that

$$
\begin{aligned}
1 &\approx \frac{V'_{av} d_{st}(\varepsilon_F)}{2} \int_0^{\alpha \beta_c \omega_D/(2\pi)-1/2} \frac{dn}{n+1/2} \\
&= \frac{V'_{av} d_{st}(\varepsilon_F)}{2} \int_{1/2}^{\alpha \beta_c \omega_D/(2\pi)-1/2+1/2} \frac{dm}{m} \\
&= \frac{V'_{av} d_{st}(\varepsilon_F)}{2} \ln(m) \Big|_{1/2}^{\alpha \beta_c \omega_D/(2\pi)} \\
&= \frac{V'_{av} d_{st}(\varepsilon_F)}{2} \left[\ln\left(\frac{\alpha \beta_c \omega_D}{2\pi}\right) - \ln\left(\frac{1}{2}\right) \right] \\
&= \frac{V'_{av} d_{st}(\varepsilon_F)}{2} \ln\left(\frac{\alpha \beta_c \omega_D}{\pi}\right)
\end{aligned} \tag{78}
$$

or

$$1 = \frac{V'_{av} d_{st}(\varepsilon_F)}{2} \ln\left(\frac{\alpha \beta_c \omega_D}{\pi}\right) \tag{79}$$

Eq. (79) may be solved using (71) broken into two pieces, instead of three, the electron–electron and the electron–phonon renormalization terms collected together, and the impurity disorder contribution:

$$V'^{RPA,CP}_{eff}(\mathbf{q}, iq_n) = W^{Coul, RPA, ren}(\mathbf{q}, iq_n) + \frac{1}{\nu} |g_{q\lambda}^{RPA, ren}|^2 D_\lambda^{RPA, ren}(\mathbf{q}, iq_n) \tag{80}$$

$$V'^{RPA, imp}_{eff} = W^{imp \; 1BA, RPA}(\mathbf{q}, iq_n) \tag{81}$$

to make the total as

$$V'^{RPA, ren}_{eff} = V'^{RPA, CP}_{eff} + V'^{RPA}_{imp} \tag{82}$$

[Notice that $V'^{RPA,CP}_{eff}$ of (80) is just the $V'^{RPA,CP}_{eff}$ found in (29).] When this decomposition is done, the solution to (79) may be written as

$$T_c = T_c^{\alpha}\Big|_{\substack{BCS\\pre}} \left(T_c\Big|_{\substack{BCS\\exp}}\right)^{\gamma_{rp}} \tag{83}$$

where the following definitions are used:

$$T_c^{\alpha}\Big|_{\substack{BCS\\pre}} = \frac{\alpha}{\pi}\frac{\hbar\omega_D}{k_B}; \quad T_c\Big|_{\substack{BCS\\exp}} = e^{-\frac{2}{V'^{RPA,CP}_{eff,av}}\frac{1}{d(\varepsilon_F)}}; \quad \gamma_{rp} = \frac{1}{1 + V'^{RPA}_{imp,av}/V'^{RPA,CP}_{eff,av}} \tag{84}$$

Here BCS as a subscript serves to identify the BCS like behavior in the prefactor coefficient and the exponential terms, whereas γ_{rp} is the exponential modification power. Subscripts "av" on the renormalized potential energies indicate the following averaging process:

$$V'^{RPA}_{eff,av} = \langle V'^{RPA}_{eff}(\mathbf{q}, iq_n; T)\big|_{iq_n \to \omega + i\eta}\rangle_{av} = V'^{RPA}_{eff}(\mathbf{q}_{av}, \omega_{av}; T) \tag{85}$$

$$V'^{RPA,CP}_{eff,av} = \langle V'^{RPA,CP}_{eff}(\mathbf{q}, iq_n; T)\big|_{iq_n \to \omega + i\eta}\rangle_{av} = V'^{RPA,CP}_{eff}(\mathbf{q}_{av}, \omega_{av}; T) \tag{86a}$$

$$V'^{RPA}_{imp,av} = \langle V'^{RPA}_{imp}(\mathbf{q}, iq_n; T)\big|_{iq_n \to \omega + i\eta}\rangle_{av} = V'^{RPA}_{imp}(\mathbf{q}_{av}, \omega_{av}; T) \tag{86b}$$

where analytic continuation is used.

By (72), (81), and (86), the γ_{rp} may be transformed into an expression that implicitly shows the impurity or disorder density concentration n_{imp},

$$\gamma_{rp} = \frac{1}{1 + \tilde{R}^V_{s,imp;CP}(n_{imp})} \tag{87}$$

where $\tilde{R}^V_{s,imp;CP}$ is the impurity to electron–phonon renormalized potential energy ratio, and in general may be expressed as

$$\tilde{R}^V_{s,imp;CP}(n_{imp}) = \tilde{R}^{V(0)}_{s,imp;CP}(0) + \tilde{R}^{V(1)}_{s,imp;CP}(0)n_{imp} + \tilde{R}^{V(2)}_{s,imp;CP}(0)\{n_{imp}\}^2 + \cdots \tag{88}$$

Numbered superscripts (i) indicate the ith derivative taken. Inserting (88) into (87) gives

$$\gamma_{rp} = \frac{1}{1 + n_{imp}[\tilde{R}^{V(0)}_{s,imp;CP}(0)\{1/n_{imp}\} + \tilde{R}^{V(1)}_{s,imp;CP}(0) + \tilde{R}^{V(2)}_{s,imp;CP}(0)n_{imp} + \cdots]} \tag{89}$$

by factoring out a power of n_{imp}, suggested by (72). Clearly, the first term within brackets contains an unphysical singularity, and requires

$$\tilde{R}^{V(0)}_{s,imp;CP}(0) = 0 \tag{90}$$

giving

$$\gamma_{rp} \approx \frac{1}{1 + n_{imp}\tilde{R}^{V(1)}_{s,imp;CP}(0)} = \frac{1}{1 + n_{imp}R^{V}_{s,imp;CP}} \tag{91}$$

The next term, $\tilde{R}^{V(1)}_{s,imp;CP}(0) = R^{V}_{s,imp;CP}$, is the impurity to electron–phonon renormalized potential energy ratio per impurity scatterer, and may or may not differ from zero, depending upon details of the nanoscopic preparation of the impurity constituents. Eq. (91) may be most consistent with non-alloying process of material preparation. Examination of the form of (91) shows that if the ratio is positive, with the impurity concentration always positive, the power modification exponential will be less than unity, and applied to the $T_c|_{BCS}^{exp}$ factor, which is a unitless number less than one, since it scales down the temperature prefactor $T_c^{\alpha}|_{BCS}^{pre}$ which is in units of Kelvin, the new effective BCS exponential will be increased in size. This will have the effect of increasing T_c. On the other hand, if the ratio is negative, the power modification exponential will be greater than unity, and when applied to the $T_c|_{BCS}^{exp}$ factor, which is a unitless number less than one, since it scales down the temperature prefactor $T_c^{\alpha}|_{BCS}^{pre}$, the new effective BCS exponential will decreased in size. This will have the effect of decreasing T_c.

If the first two terms in the brackets of (89) are zero, as some interpretations seem to have favored or implied previously [see Abrikosov and Gor'kov (1961) which states "It is well known that the transition temperature of superconductors with nonmagnetic impurities remains practically unchanged in the region of low concentrations", which cites Abrikosov and Gor'kov (1959b) treating alloys at $T = 0$, and Abrikosov and Gor'kov (1959a) treating alloys at $T \neq 0$, although the latter two references do not seem explicitly to prove that assertion; or Tinkham (1980) on page 263 stating "when we considered Anderson's theory of dirty superconductors (Anderson, 1959), i.e., nonmagnetic alloys with mean free path $\ell < \xi_0$, we note that pairing of time-reversed degenerate states led to same T_c and BCS density of states as for a pure superconductor"; where Anderson (1959) made qualitative arguments for disorder effects based upon some first and second order quantum mechanical perturbation theory], creating a stronger null as $n_{imp} \to 0$, that is

$$\tilde{R}^{V(0)}_{s,imp;CP}(0) = 0; \qquad \tilde{R}^{V(1)}_{s,imp;CP}(0) = 0 \tag{92}$$

then

$$\gamma_{rp} \approx \frac{1}{1 + (n_{imp})^2 \tilde{R}^{V(2)}_{s,imp;CP}(0)} \tag{93}$$

Eq. (93) may be most consistent with alloy processes. Note that both γ_{rp} forms (91) and (93) do satisfy the necessary condition

$$\lim_{n_{imp} \to 0} \gamma_{rp} = 1 \tag{94}$$

In order to have a weaker march to unity in (94), we will address the case of (91) below.

There is another subtlety to interpreting (91), that is, increasing the number of impurities may reduce the disorder in the material when constructively affecting the arrangement of the atomic lattice, so the actual disorder number, n_{dis}, will no longer equal n_{imp}, requiring (91) be modified to

$$\gamma_{rp} = \frac{1}{1 + n_{dis} R^{V}_{s,imp;CP}} \tag{95}$$

In this case, and relative to a nominal value, n_{dis} may go negative, and the power modification exponential, will increase, if $R^{V}_{s,imp;CP}$ is positive. As a consequence, the critical temperature T_c will decrease. However, if $R^{V}_{s,imp;CP}$ is negative, critical temperature T_c will increase.

When modifying second term in the (95) denominator is considerably smaller than unity, (83), (84), and (95) allow T_c to be written as

$$T_c = T_c^{BCS}(T_c|_{\substack{BCS \\ \exp}})^{-n_{dis}R}; \qquad T_c^{BCS} = \frac{\alpha}{\pi} \frac{\hbar \omega_D}{k_B} T_c|_{\substack{BCS \\ \exp}};$$

$$T_c|_{\substack{BCS \\ \exp}} = e^{-\frac{2}{V_{eff,av}^{RPA,CP}} \frac{1}{d(\varepsilon_F)}} \tag{96}$$

The result in (96) is sketched in Fig. 1, where R stands for $R^{V}_{s,imp;CP}$ which is the impurity to electron–phonon renormalized potential energy ratio, and T_c is plotted against n_{dis}. Notice, negative going values of n_{dis} increase T_c whereas positive going values decrease T_c. Parametrization in terms of

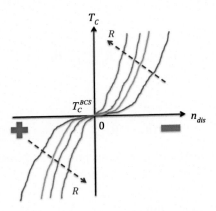

Figure 1 Critical superconducting temperature T_c versus disorder density n_{dis}, parameterized in terms of electron–phonon renormalized energy ratio R.

R is done. Note that in no cases may $n_{dis} < 0$ be used to generate a superconducting effect when none existed in the material ($n_{dis} = 0$ prior to disturbing the material), as that would be unphysical.

One may obtain a linearized version of (96) using a Taylor's expansion, such that

$$T_c(n_{dis}) = T_c^{(0)} + T_c^{(1)} n_{dis} + O([n_{dis}]^2) \tag{97}$$

When this is done, $T_c^{(0)}$ and $T_c^{(1)}$ are determined to be

$$T_c^{(0)} = \frac{\alpha}{\pi} \frac{\hbar \omega_c}{k_B} e^{-\frac{1}{V_{eff,av}^{\prime RPA,CP} d(\varepsilon_F)/2}}, \qquad T_c^{(1)} = T_c^{(0)} \frac{2R_{s,imp;CP}^V}{V_{eff,av}^{\prime RPA,CP} d(\varepsilon_F)} \tag{98}$$

Appearance of a finite $T_c^{(1)}$ coefficient is consistent with using a form of the Eliashberg equation employing the phonon spectral density $\alpha^2(\omega)F(\omega)$, finding additional low frequency transverse while reduced longitudinal electron–phonon coupling, giving a net effective gain in weighting by the phonon spectral density. This yields a critical temperature shift of $(T_c - T_c^p)/T_c^p = C_p/(q_D \ell_{mfp})$ (Keck & Schmid, 1976). Here q_D is the Debye momentum (related to Debye temperature), ℓ_{mfp} the mean free path of electrons, and T_c^p the critical temperature for pure metals, with T_c the critical temperature for non-magnetic impurities. Arguing that $\ell_{mfp} \propto 1/n_{imp}$ yields $T_c^{(1)} \propto n_{imp}$.

12. RELATING THE DISORDER POTENTIAL ENERGY TO THE GAP PARAMETER

The gap parameter used in the previous sections is not the regular or ordinary gap parameter. Rather, it has been modified by the impurity scattering potential energy, and so that altered gap is derived here. There are less typical or expected quantum many-body Green's functions, often referred to as anomalous Green's functions. One of them which will be of use is defined as

$$F_{\mathbf{k}\downarrow\uparrow}(\mathbf{k}, \tau) = -\langle T_\tau \{ c^\dagger_{-\mathbf{k}\downarrow}(\tau) c^\dagger_{\mathbf{k}\uparrow}(0) \} \rangle \tag{99}$$

T_τ is the time ordering operator in the imaginary Matsubara time τ frame. The bracketed operation indicates an ensemble statistical average over any complete eigenstate set ν. Thus for some operator O,

$$\langle O \rangle = \frac{1}{Z} \sum_\nu \langle \nu | O | \nu \rangle e^{-\beta E_\nu}; \quad Z = \sum_\nu e^{-\beta E_\nu} \tag{100}$$

Here Z is the partition function, formed by the exponential sum using the system Hamiltonian eigenenergies E_ν. The signature property of the anomalous Green's function is that it does not have mixed raising and lowering operators as the ordinary Green's function

$$G_{\uparrow\uparrow}(\mathbf{k}, \tau) = -\langle T_\tau \{ c_{\mathbf{k}\uparrow}(\tau) c^\dagger_{\mathbf{k}\uparrow}(0) \} \rangle \tag{101}$$

These two types of Green's functions can be related by equations of motion,

$$\partial_\tau G_{\uparrow\uparrow}(\mathbf{k}, \tau) = -\delta(\tau) - \xi_\mathbf{k} G_{\uparrow\uparrow}(\mathbf{k}, \tau) + \Delta_\mathbf{k} F_{\downarrow\uparrow}(\mathbf{k}, \tau) \tag{102a}$$

$$\partial_\tau F_{\downarrow\uparrow}(\mathbf{k}, \tau) = \qquad -\xi_\mathbf{k} F_{\downarrow\uparrow}(\mathbf{k}, \tau) + \Delta_\mathbf{k} G_{\uparrow\uparrow}(\mathbf{k}, \tau) \tag{102b}$$

It is worth noting that the 2nd quantized raising and lowering electron operators employed in (99) and (101) are perturbed many-body operators.

A mean-field approach relates the total potential energy to the gap parameter using 2nd quantized electron operators. From it we need to find actual relations which explicitly demonstrate the relationship between the disorder effect via the impurity scattering potential energy, to the gap parameter. The starting point is recognizing that the expression uses the anomalous Green's function, converting into the Matsubara frequency domain, retrieving the 4-momentum space formula for it, eventually performing an integration.

$$
\begin{aligned}
\Delta_{\mathbf{k}} &= \sum_{\mathbf{k}'}^{|\xi_{\mathbf{k}'}|<\omega_D} V'_{\mathbf{k}\mathbf{k}'} \langle c_{-\mathbf{k}'\downarrow} c_{\mathbf{k}\uparrow} \rangle \\
&\approx V'^{RPA}_{eff,av} \sum_{\mathbf{k}'}^{|\xi_{\mathbf{k}'}|<\omega_D} F^*_{\downarrow\uparrow}(\mathbf{k},\tau)\big|_{\tau\to 0} \\
&= V'^{RPA}_{eff,av} \sum_{\mathbf{k}}^{|\xi_{\mathbf{k}'}|<\omega_D} \sum_{ik_n} e^{-ik_n\tau} F^*_{\downarrow\uparrow}(\mathbf{k},ik_n)\big|_{\tau\to 0} \\
&= V'^{RPA}_{eff,av} \sum_{\mathbf{k}}^{|\xi_{\mathbf{k}'}|<\omega_D} \sum_{ik_n} e^{-ik_n\cdot 0^+} F^*_{\downarrow\uparrow}(\mathbf{k},ik_n) \\
&= V'^{RPA}_{eff,av} \sum_{\mathbf{k}}^{|\xi_{\mathbf{k}'}|<\omega_D} \sum_{ik_n} e^{-ik_n\cdot 0^+} \left[\frac{-\Delta^*_{\mathbf{k}}}{(ik_n)^2-(E_{\mathbf{k}})^2}\right]^* \\
&= -V'^{RPA}_{eff,av} \sum_{\mathbf{k}}^{|\xi_{\mathbf{k}}|<\omega_D} \Delta_{\mathbf{k}} \frac{1}{\beta} \sum_{ik_n} e^{-ik_n\cdot 0^+} \frac{1}{(ik_n)^2-(E_{\mathbf{k}})^2} \\
&= -V'^{RPA}_{eff,av} \sum_{\mathbf{k}}^{|\xi_{\mathbf{k}'}|<\omega_D} \Delta_{\mathbf{k}} \frac{1}{\beta} \sum_{ik_n} e^{-ik_n\cdot 0^+} \frac{1}{(ik_n-|E_{\mathbf{k}}|)(ik_n+|E_{\mathbf{k}}|)} \quad (103)
\end{aligned}
$$

$\Delta_{\mathbf{k}}$ is readily evaluated by using the contour integration property for the complex valued function $g_0(z)$ with poles in the complex z-plane,

$$
\frac{1}{\beta} \sum_{ik_n} g_0(ik_n) e^{ik_n\tau}\bigg|_{\tau>0} = \sum_{j=1}^{M} f(z_j) e^{z_j\tau} \operatorname*{Res}_{z=z_j}[g_0(z)] \quad (104)
$$

where f is just the Fermi–Dirac function. Noting in (103) we used the following for $F_{\downarrow\uparrow}(\mathbf{k},\tau)$ and identifying $g_0(z)$ as

$$
F_{\downarrow\uparrow} = \frac{-\Delta^*_{\mathbf{k}}}{(ik_n)^2-(E_{\mathbf{k}})^2}; \qquad g_0(z) = \frac{1}{(z-|E_{\mathbf{k}}|)(z+|E_{\mathbf{k}}|)} \quad (105)
$$

one finds

$$
\Delta_{\mathbf{k}} = -V'^{RPA}_{eff,av} \sum_{\mathbf{k}}^{|\xi_{\mathbf{k}'}|<\omega_D} \Delta_{\mathbf{k}} \frac{f(\beta E^+_{\mathbf{k}})-f(\beta E^-_{\mathbf{k}})}{2E^+_{\mathbf{k}}}
$$

$$= -V'^{RPA}_{eff,av} \sum_{\mathbf{k}}^{|\xi_{\mathbf{k}'}|<\omega_D} \Delta_{\mathbf{k}} \frac{f(\beta|E_{\mathbf{k}}|) - f(-\beta|E_{\mathbf{k}}|)}{2|E_{\mathbf{k}}|}$$

$$= -V'^{RPA}_{eff,av} \sum_{\mathbf{k}}^{|\xi_{\mathbf{k}'}|<\omega_D} \Delta_{\mathbf{k}} \frac{f(\beta E_{\mathbf{k}}) - f(-\beta E_{\mathbf{k}})}{2E_{\mathbf{k}}}$$

$$= V'^{RPA}_{eff,av} \sum_{\mathbf{k}}^{|\xi_{\mathbf{k}'}|<\omega_D} \Delta_{\mathbf{k}} \frac{1 - 2f(\beta E_{\mathbf{k}})}{2E_{\mathbf{k}}} \tag{106}$$

where we dropped the magnitude signs on the dispersion relation, taking by default the positive branch.

Considering the simplest condition where the gap parameter becomes momentum independent, $\Delta_{\mathbf{k}} \to \Delta$, will factor out of the summation, leaving it to be divided out,

$$1 = V'^{RPA}_{eff,av} \sum_{\mathbf{k}}^{|\xi_{\mathbf{k}'}|<\omega_D} \frac{1 - 2f(\beta E_{\mathbf{k}})}{2E_{\mathbf{k}}}$$

$$= V'^{RPA}_{eff,av} \int_{-\varepsilon_F}^{\infty} D(\xi_{\mathbf{k}}) \frac{1 - 2f(\beta E_{\mathbf{k}})}{2E_{\mathbf{k}}} d\xi_{\mathbf{k}}$$

$$\approx V'^{RPA}_{eff,av} D(\varepsilon_F) \int_{-\varepsilon_F}^{\infty} \frac{1 - 2f(\beta E_{\mathbf{k}})}{2E_{\mathbf{k}}} d\xi_{\mathbf{k}}$$

$$\approx V'^{RPA}_{eff,av} D(\varepsilon_F) \int_{-\omega_D}^{\omega_D} \frac{1 - 2f(\beta E_{\mathbf{k}})}{2E_{\mathbf{k}}} d\xi_{\mathbf{k}}$$

$$\approx V'^{RPA}_{eff,av} D(\varepsilon_F) \int_{-\omega_D}^{\omega_D} \frac{\tanh(\frac{\beta}{2}\sqrt{(\xi_{\mathbf{k}})^2 + |\Delta|^2})}{2\sqrt{(\xi_{\mathbf{k}})^2 + |\Delta|^2}} d\xi_{\mathbf{k}} \tag{107}$$

The final equality in (107) provides the relationship between the average effective RPA potential energy, as given by (85) and (86), and the gap parameter. [Note that $d(\varepsilon_F) = d_{st}(\varepsilon_F) = D(\varepsilon_F)/\mathcal{V}$, where $\mathcal{V}$ is the volume in 3D, which is replaced by area A_{2D} in 2D.]

A non-transcendental relationship is found as $T \to 0$,

$$|\Delta(0)| \approx |\Delta(0)|_{\substack{BCS \\ pre}} (|\Delta(0)|_{\substack{BCS \\ exp}})^{\gamma_{rp}} \tag{108}$$

where

$$|\Delta(0)|_{\substack{BCS \\ pre}} = 2\hbar\omega_D; \qquad |\Delta(0)|_{\substack{BCS \\ exp}} = e^{-\frac{1}{V'^{RPA,CP}_{eff,av} D(\varepsilon_F)}} \tag{109}$$

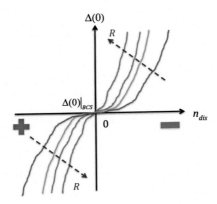

Figure 2 Gap $\Delta(0)$ at zero ambient temperature versus the disorder density n_{dis}, parameterized in terms of the electron–phonon renormalized energy ratio R.

As we saw earlier when examining the behavior of the critical temperature T_c, with a positive ratio, the power modification exponential will be less than unity, and applied to the $|\Delta(0)|_{BCS\ exp}$ factor, which is a unitless number less than one, the new effective BCS exponential will increased in size. This will have the effect of increasing $|\Delta(0)|$. On the other hand, if the ratio is negative, the power modification exponential will be greater than unity, and when applied to the $|\Delta(0)|_{BCS\ exp}$ factor, which is a unitless number less than one, since it scales down the temperature prefactor $T_c^\alpha|_{BCS\ pre}$, the new effective BCS exponential will decreased in size. This will have the effect of decreasing $|\Delta(0)|$.

There is another subtlety to interpreting (87), that is, increasing the number of impurities may reduce the disorder in the material when constructively affecting the arrangement of the atomic lattice, so the actual disorder number, n_{dis}, will no longer equal n_{imp}. In this case and relative to a nominal value, n_{dis} may go negative, and (91) changes into (95), making $\gamma_{rp} = 1/[1 + n_{dis}R_{s,imp;CP}^V]$ increase, enhancing the value of $|\Delta(0)|$.

When modifying second term in the (95) denominator is considerably smaller than unity, (83), (84), and (95) allow T_c to be written as

$$\Delta(0) = \Delta_{BCS}(0)\left(\Delta(0)\big|_{BCS\ exp}\right)^{-n_{dis}R}; \qquad \Delta_{BCS}(0) = 2\hbar\omega_D\Delta(0)\big|_{BCS\ exp};$$

$$\Delta(0)\big|_{BCS\ exp} = e^{-\frac{2}{V_{eff,av}^{RPA,CP}}\frac{1}{d(\varepsilon_F)}} \qquad (110)$$

The result in (110) is sketched in Fig. 2, where R stands for $R_{s,imp;CP}^{V}$ which is the impurity to electron–phonon renormalized potential energy ratio, and $\Delta(0)$ is plotted against n_{dis}. Notice, negative going values of n_{dis} increase $\Delta(0)$ whereas positive going values decrease $\Delta(0)$. Parametrization in terms of R is done.

13. CONCLUSIONS

Disorder as it microscopically affects superconducting properties, has been delineated in the treatment given here. Useful, compact analytical expressions for two major parameters of interest, the critical temperature T_c and the gap Δ, are found. The indirect relationship between T_c and gap parameter Δ is shown when obtaining Δ's dependence on the same disorder potential energy quantity as T_c. The treatment shown here opens up the possibility that disorder could possibly increase T_c, something not seen in the 1980s, but observed in some circumstances as mentioned previously in the Introduction here. Hints of that possibility were indicated earlier in Keck and Schmid (1976).

The results here are not inconsistent with the original BCS theory (Bardeen, Cooper, & Schrieffer, 1957), although formulas provided here are derived in a more streamlined fashion, not necessarily relying on band-structure symmetries originally utilized in either reciprocal **k**-space or spin index σ.

ACKNOWLEDGMENTS

This contribution arose while working on the project Lower Dimensional Materials for Naval Applications, involving our Electronics Science & Technology, Material Science & Technology, and Chemistry Divisions, with the Plasma Physics Division also participating as a collaborating division, at the Naval Research Laboratory, Washington, DC. Interactions with all of the project researchers in these divisions have informed the contents of the present work, over the course of the project, from October 2013 to October 2017. I mention with particular gratitude the many interesting discussions with Dr. Michael S. Osofsky of MSTD, on electron–electron interactions, superconductivity, metal–insulator transition, and many other topics related to 2D solid state material systems. Finally, I thank several unnamed researchers, whose suggestions have been incorporated in improving the manuscript, reflected also in some of the references selected.

REFERENCES

Abrahams, E., Anderson, P. W., Licciardello, D. C., & Ramakrishnan, T. V. (1979). Scaling theory of localization: Absence of quantum diffusion in two dimensions. *Physical Review Letters, 42*, 673–676.

Abrikosov, A. A. (2004). Nobel lecture: Type-II superconductors and the vortex lattice. *Reviews of Modern Physics, 76*, 975.

Abrikosov, A. A., & Gor'kov, L. P. (1959a). Superconducting alloys at finite temperatures. *Soviet Physics, JETP, 9*, 220 (in Russian, *Journal of Experimental and Theoretical Physics (USSR), 36*, 319, Jan. 1959).

Abrikosov, A. A., & Gor'kov, L. P. (1959b). On the theory of superconducting alloys 1. The electrodynamics of alloys at absolute zero. *Soviet Physics, JETP, 35*, 1090 (in Russian, *Journal of Experimental and Theoretical Physics (USSR), 35*, 1558, Dec. 1958).

Abrikosov, A. A., & Gor'kov, L. P. (1961). Contributions to the theory of superconducting alloys with paramagnetic impurities. *Soviet Physics, JETP, 12*, 1243 (in Russian, *Journal of Experimental and Theoretical Physics (USSR), 39*, 1781, Dec. 1960).

Abrikosov, A. A., Gorkov, L. P., & Dzyaloshinski, I. E. (1963). *Methods of quantum field theory in statistical physics* (R. A. Silverman, Trans. Ed.). Englewood Cliffs, NJ: Prentice-Hall, Inc.

Abrikosov, A. A., Gorkov, L. P., & Dzyaloshinski, I. Ye. (1965). *Quantum field theoretical methods in statistical physics* (D. E. Brown, Trans., D. ter Haar, Ed.). Oxford: Pergamon Press (Original ed. publ. Fizmatgiz, Moscow, 1962 & added mater. 1964).

Anderson, P. W. (1959). Theory of dirty superconductors. *Journal of Physics and Chemistry of Solids, 11*, 26–30.

Attanasi, A. (2008). *Competition between superconductivity and charge density waves: The role of disorder* (Ph.D. thesis). Sapienza Universita di Roma.

Balatsky, A. V., Vekhter, I., & Zhu, J.-X. (2006). Impurity-induced states in conventional and unconventional superconductors. *Reviews of Modern Physics, 78*, 373.

Bardeen, J., Cooper, L. N., & Schrieffer, J. R. (1957). Theory of superconductivity. *Physical Review, 108*, 1175.

Bartolf, H. (2016). *Fluctuation mechanisms in superconductors.* Springer Spektrum.

Bellafi, B., Haddad, S., & Charfi-Kaddour, S. (2009). Disorder-induced superconductivity in ropes of carbon nanotubes. *Physical Review B, 80*, 075401.

Bonetto, C., Israeloff, N. E., Pokrovskiy, N., & Bojko, R. (1998). Field induced superconductivity in disordered wire networks. *Physical Review, 58*, 128.

Bruus, H., & Flensberg, K. (2004). *Many-body quantum theory in condensed matter physics – An introduction.* Oxford University Press (reprinted in 2013).

Bugoslavsky, Y., Cohen, L. F., Perkins, G. K., Polichetti, M., Tate, T. J., Gwilliam, R., et al. (2001). Enhancement of the high-magnetic-field critical current density of superconducting MgB_2 by proton irradiation. *Nature, 411*, 561.

Corbette, J. P. (1990). Properties of boson-exchange superconductors. *Reviews of Modern Physics, 62*, 1027.

Das Sarma, S., Adam, S., Hwanf, E. H., & Rossi, E. (2011). Electronic transport in two-dimensional graphene. *Reviews of Modern Physics, 83*, 407–470.

Driessen, E. F. C., Coumou, P. C. J. J., Tromp, R. R., de Visser, P. J., & Klapwijk, T. M. (2012). Strongly disordered TiN and NbTiN s-wave superconductors probed by microwave electrodynamics. *Physical Review Letters, 109*, 107003.

Eliashberg, G. M. (1960). Interactions between electrons and lattice vibrations in a supercon-
ductor. *Soviet Physics, JETP, 11*, 696 (in Russian, *Journal of Experimental and Theoretical Physics (USSR), 38*, 966, Mar. 1960).

Eliashberg, G. M. (1961). Temperature Green's function for electrons in a superconductor. *Soviet Physics, JETP, 12*, 1000 (in Russian, *Journal of Experimental and Theoretical Physics (USSR), 39*, 1437, Nov. 1960).

Fetter, A. L., & Walecka, J. D. (1971). *Quantum theory of many-particle systems*. McGraw-Hill.

Finkelstein, A. M. (1987). Superconducting transition temperature in amorphous films. *Pis'ma v Zhurnal Eksperimental'noi i Teoreticheskoi Fiziki, 45*, 37. *JETP Letters, 45*, 46.

Fisher, D. S., Fisher, M. P. A., & Huse, D. A. (1991). Thermal fluctuations, quenched disorder, phase transitions, and transport in type-II superconductors. *Physical Review B, 43*, 130.

Galitski, V. M., & Larkin, A. I. (2001). Disorder and quantum fluctuations in superconducting films in strong magnetic fields. *Physical Review Letters, 87*, 087001.

Ginzburg, V. L., & Kirzhnits, D. A. (Eds.). (1982). *High-temperature superconductivity*. New York: Consultants Bureau (A. K. Agyer, Trans., J. L. Birman, Ed.; original Russian text, Nauka, Moscow, 1977).

Karnaukhov, I. M., & Shepelev, A. G. (2008). Type II superconductors are 70 years old. *Europhysics News, 39*, 35.

Keck, B., & Schmid, A. (1976). Superconductivity and electron–phonon interaction in impure metals. *Journal of Low Temperature Physics, 24*, 611.

Kemper, A., Doluweera, D. G. S. P., Maier, T. A., Jarrell, M., Hirschfeld, P. J., & Cheng, H-P. (2009). Insensitivity of superconductivity to disorder in the cuprates. Retrieved from arXiv:0807.0195v2.

Kierfeld, J., & Vinokur, V. (2004). Lindemann criterion and vortex lattice phase transitions in type-II superconductors. *Physical Review B, 69*, 024501.

Kim, J. H., Dou, S. X., Oh, S., Jercinovic, M., Babic, E., Nakane, T., et al. (2008). Correla-
tion between doping induced disorder and superconducting properties in carbohydrate doped MgB_2. Retrieved from arXiv:0810.1558.

Kittel, C. (1987). *Quantum theory of solids*. New York: Wiley.

Kotliar, G., & Kapitulnik, A. (1986). Anderson localization and the theory of dirty superconductors. II. *Physical Review B, 33*, 3146.

Kozhevnikov, V. F., Van Bael, M. J., Vinckx, W., Temst, K., Van Haesendonck, C., & Indekeu, J. O. (2005). Surface enhancement of superconductivity in tin. *Physical Review B, 72*, 174510.

Krowne, C. M. (2011). Nanowire and nanocable intrinsic quantum capacitances and junc-
tion capacitances: Results for metal and semiconducting oxides. *Journal of Nanomaterials*, 160639.

Maleyev, S. V., & Toperverg, B. P. (1988). On high-T_c superconductivity and structural disorder. *Solid State Communications, 67*, 405.

Martinez, J. I., Abad, E., Calle-Vallejo, F., Krowne, C. M., & Alonso, J. A. (2013). Tailor-
ing structural and electronic properties of RuO_2 nanotubes: Many-body approach and electronic transport. *Physical Chemistry Chemical Physics, 15*(35), 14715–14722.

Martinez, J. I., Calle-Vallejo, F., Krowne, C. M., & Alonso, J. A. (2012). First-principles structural & electronic characterization of ordered SiO_2 nanowires. *The Journal of Physical Chemistry C, 116*, 18973–18982.

Mondaini, F., Paiva, T., dos Santos, R. R., & Scalettar, R. T. (2008). Disordered two-
dimensional superconductors: Role of temperature and interaction strength. *Physical Review B, 78*, 174519.

Nakhmedov, E., Alekperov, O., & Oppermann, R. (2012). Effects of randomness on the critical temperature in quasi-two-dimensional organic superconductors. *Physical Review B, 86*, 214513.

Osofsky, M. S., Hernández-Hangarter, S. C., Nath, A., Wheeler, V. D., Walton, S. G., Krowne, C. M., et al. (2016). Functionalized graphene as a model system for the two-dimensional metal–insulator transition. *Scientific Reports, 6*, 19939.

Osofsky, M. S., Krowne, C. M., Charipar, K. M., Bussmann, K., Chervin, C. N., Pala, I. R., et al. (2016). Disordered RuO_2 exhibits two dimensional, low-mobility transport and a metal–insulator transition. *Scientific Reports, 6*, 21836.

Pines, D. (1964). *Elementary excitations in solids: Lectures on phonons, electrons and plasmons*. New York: W. A. Benjamin, Inc. (First publ. 1963; corrections in 1964, 2nd printing).

Pines, D. (1979). *The many-body problem* (5th printing). *Frontiers in physics, a lecture note & reprint volume*. Reading, MA: Benjamin/Cummings Publ. Co.

Schrieffer, J. R. (1964). *Theory of superconductivity. Frontiers in physics*. Reading, MA: Perseus Books (Nobel lectures are available for the BCS theory in the 1983 edition).

Siemons, W., Steiner, M. A., Koster, G., Blank, D. H. A., Beasley, M. R., & Kapitulnik, A. (2008). Preparation and properties of amorphous MgB_2/MgO superstructures: Model disordered superconductor. *Physical Review B, 77*, 174506.

Su, X., Zuo, F., Schlueter, J. A., Kelly, M. E., & Williams, J. M. (1998). Structural disorder and its effect on the superconducting transition temperature in the organic superconductor κ-$(BEDT\text{-}TTF)_2Cu[N(CN)_2]Br$. *Physical Review B, 57*, R14056.

Swanson, M., Loh, Y. L., Randeria, M., & Trivedi, N. (2014). Dynamical conductivity across the disorder-tuned superconductor–insulator transition. *Physical Review X, 4*, 021007.

Tanaka, K., & Marsiglio, F. (2000a). Possible electronic shell structure of nanoscale superconductors. *Physics Letters A, 265*, 133.

Tanaka, K., & Marsiglio, F. (2000b). Anderson prescription for surfaces and impurities. *Physical Review B, 62*, 5345.

Tinkham, M. (1980). *Introduction to superconductivity*. Malabar, FL: Krieger Publ. Co. (Orig. ed., McGraw-Hill, 1975).

Wang, Y. L., Wu, X. L., Chen, C.-C., & Lieber, C. M. (1990). Enhancement of the critical current density in single-crystal $Bi_2Sr_2CaCu_2O_8$ superconductors by chemically induced disorder. *Proceedings of the National Academy of Sciences of the United States of America, 87*, 7058.

CHAPTER THREE

The Struggle to Overcome Spherical Aberration in Electron Optics

Albert Septier✠

Institute of Electronics, Faculty of Sciences of Orsay, University of Paris, Orsay, France

Contents

✠ Deceased. "Reprinted from A. Septier (1966). Advances in Optical and Electron Microscopy, vol. 1, pp. 204–274 (Academic Press, London & New York)".

Advances in Imaging and Electron Physics, Volume 202
ISSN 1076-5670
http://dx.doi.org/10.1016/bs.aiep.2017.06.001

1. INTRODUCTION

In electron optics as in classical optics, the aperture aberration is by far the most important, for it is this aberration which limits the resolution of electron microscopes and the smallness of the probes of microanalyzers. It is for this reason that the battle against this aberration has been (and continues to be) waged with such tenacity.

The first treatise on electron optics, written by Brüche and Scherzer, was published in 1934; since that date, considerable efforts have been devoted to the problem of reducing or correcting the spherical aberration of lenses, both in theory and in practice. Although Scherzer had shown in 1936 (Scherzer, 1936a) that electron optics differs from classical optics in that the spherical aberration of ordinary rotationally symmetrical systems cannot be corrected, hope nevertheless persisted for several years that it might be possible to circumvent this difficulty (Glaser, 1940); in 1940, however, Recknagel demonstrated incontrovertibly that these attempts could never succeed.

Until 1947, all efforts were directed at obtaining lenses with little aberration, by seeking the optimum operating conditions of any given lens, and the field and potential distributions which would correspond to minimum aberration. However, after the publication of another article by Scherzer (1947), in which he analyzed the various possible methods of invalidating the result which he had proved in 1936, namely, that correction was impossible, a new approach became apparent, and the problem was reconsidered in many articles. At the same time, lenses continued to be improved, and today all electron optical instruments employ lenses with little spherical aberration, until such time as a correction system becomes available which is simple enough in design and adjustment.

After recapitulating the expression for the coefficient of spherical aberration, we shall first describe the attempts which have been made to improve round lenses, and then review the different methods of achieving complete correction; finally, we shall discuss the present state of the problem.

2. THE COEFFICIENT OF SPHERICAL ABERRATION

2.1 Definition

Let A be a point object, lying on the axis Oz of a centered system L, and let P be the corresponding image plane of Gaussian optics (the incident rays

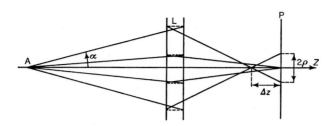

Figure 1 Longitudinal and transverse aperture defects of a lens. P = Gaussian image plane, L = thick lens.

are only slightly inclined to the axis: $\alpha < 10^{-2}$ radians). When α is large (Fig. 1), the rays intersect the axis between L and P, and for a conical beam of semi-aperture α, a circular aberration patch of radius ρ will be seen at P. If M denotes the linear magnification of the system L in this situation, and f its focal length, we can express ρ in the form

$$\rho = M(C_3\alpha^3 + C_5\alpha^5 + \ldots) \tag{1}$$

or

$$\rho = M(C_s f\alpha^3 + C_s' f\alpha^5 + \ldots). \tag{2}$$

C_3 and C_s are the spherical aberration coefficients of the lens, to the third order approximation, and C_5, C_s' the fifth order coefficients; under normal working conditions, the first aberration term predominates, and the fifth order terms are very often neglected. If C_3 could be reduced to a small enough value, or to zero, this would of course no longer be permissible.

The coefficients C_3 (and C_s) can be obtained straightforwardly by experiment. The expressions which give them as functions of the potentials or fields are, on the contrary, extremely complicated; they can be calculated rapidly only in the very rare cases when the functions $\Phi(z)$ and $B(z)$, which describe the axial distributions of potential and magnetic field, can be represented by suitable analytical expressions.

The coefficients can be obtained by a simple calculation, once the first order (or Gaussian) equations of motion have been solved; the aberration is regarded as a small perturbation of the Gaussian trajectory due to the higher order terms in the expansions of $\Phi(z)$ and $B(z)$ in the neighborhood of the axis.

2.2 The General Expressions for the Coefficients

In the electrostatic case, we write $\Phi'(z)/\Phi(z) = u$ and we denote by $r_\alpha(z)$ the particular trajectory for which $r_\alpha(z_0) = 0$ and $r'_\alpha(z_0) = 1$; we find (Glaser, 1952)

$$C_3 = -\frac{1}{64\Phi_0^{\frac{1}{2}}} \int_{z_0}^{z_i} \Phi^{\frac{1}{2}}(z)\left(4u'^2 + 3u^4 - 5u^2u' - uu''\right)r_\alpha^4 dz. \tag{3}$$

z_0 and z_i denote the abscissae of the object and its image, respectively; the symbols Φ', r', u' ... denote the derivatives of Φ, r, u... with respect to the variable z.

Likewise, in the magnetic case, we find

$$C_3 = \frac{1}{96} \cdot \frac{e}{m_0\Phi_0^*} \int_{z_0}^{z_i} \left(\frac{2e}{m_0\Phi_0^*}B^4(z) + 5B'^2(z) - B(z)B''(z)\right)r_\alpha^4 dz. \tag{4}$$

Φ_0^* is now the accelerating potential, relativistically corrected if necessary; for electrons, the formula for this correction can be written

$$\Phi_0^* = \Phi_0\left(1 + 10^{-6}\Phi_0\right).$$

The higher order derivatives are difficult to obtain from experiment; they can, however, be eliminated, and we obtain the following expressions:

$$C_3 = \frac{1}{64\Phi_0^{\frac{1}{2}}} \int_{z_0}^{z_i} \Phi^{\frac{1}{2}}(z)\left[\left(3u^4 + \frac{9}{2}u^2u' + 5u'^2\right)r_\alpha^4 + 4uu'r_\alpha^3 r'_\alpha\right]dz \tag{5}$$

and

$$C_3 = \frac{1}{128} \cdot \frac{e}{m_0\Phi_0^*} \int_{z_0}^{z_i} \left[\left(\frac{3e}{m_0\Phi_0^*}B^4(z) + 8B'^2(z)\right)r_\alpha^4 - 8B^2(z)r_\alpha^2 r'^2_\alpha\right]dz. \tag{6}$$

r_α is a solution of the first-order equation:

$$\frac{d}{dz}\left(\Phi^{\frac{1}{2}}(z)\frac{dr}{dz}\right) + \frac{r}{4\Phi^{\frac{1}{2}}(z)}\Phi''(z) = 0 \tag{7}$$

or

$$\frac{d^2r}{dz^2} + \frac{e}{8m_0\Phi_0^*} \cdot B^2(z)r = 0. \tag{8}$$

2.3 The Impossibility of Correction in a Centered System Without Space Charge

In 1936, Scherzer showed that the expressions given above can be cast into the form of a sum of squares, all with the same sign, provided that there is no free electric charge within the lens. In the general case (Scherzer, 1936a), where $\Phi(z)$ and $B(z)$ are superimposed, the expression for C_3 becomes

$$
C_3 = \frac{1}{16\Phi_0^{\frac{1}{2}}} \int_{z_0}^{z_i} \Phi^{\frac{1}{2}} \left[\frac{5}{4}\left(\frac{\Phi''}{\Phi} + \frac{\Phi'}{\Phi}\frac{r'_\alpha}{r_\alpha} - \frac{\Phi'^2}{\Phi^2} \right)^2 + \left(\frac{\Phi'}{\Phi} \right)^2 \left(\frac{r'_\alpha}{r_\alpha} + \frac{7}{8}\frac{\Phi'}{\Phi} \right)^2 \cdots \right.
$$
$$
+ \frac{e}{m_0\Phi}\left(B' + B\frac{r'_\alpha}{r_\alpha} - \frac{5}{4}B\frac{\Phi'}{\Phi} \right)^2 + \frac{e}{m_0} \cdot \frac{B^2}{\Phi}\left(\frac{r'_\alpha}{r_\alpha} + \frac{1}{4}\frac{\Phi'}{\Phi} \right)^2
$$
$$
\left. + \frac{1}{64}\frac{\Phi'^4}{\Phi^4} + \frac{e^2}{4m_0^2}\frac{B^4}{\Phi^4} + \frac{e}{32m_0}B^2\frac{\Phi'^2}{\Phi^3} \right] r_\alpha^4 dz. \tag{9}
$$

When $\Phi(z) = \Phi_0$, we have the case of magnetic lenses alone, and when $B(z) = 0$, of electrostatic lenses.

This relation demonstrates irrefutably that C_3 can never be reduced to zero by a judicious choice of the distributions $\Phi(z)$ and $B(z)$. A careful scrutiny of the method by which this expression is obtained reveals that it is valid only if the following conditions are satisfied:

(a) the optical system must have rotational symmetry;
(b) an object placed at z_0 produces a real image at z_i;
(c) the potentials are static: they have no time variation;
(d) there is no space charge;
(e) the functions $\Phi(z)$ and $\Phi'(z)/\Phi(z)$ are not discontinuous anywhere along the axis Oz.

We shall consider the possible ways of designing corrector systems by exploiting these conditions in a later section.

2.4 The Round Lens "Without Aberration"

At one time, Glaser (1940) believed that it would be possible to build a magnetic lens free of aberration. Setting out from Eq. (4), he solved the equation

$$
\int_{z_0}^{z_i} \left[\frac{2e}{m_0\Phi_0^*}B^4(z) + 5B'^2(z) - B'(z)B''(z) \right] r_\alpha^4 \, dz = 0 \tag{10}
$$

and showed that a distribution $B(z)$ which satisfies (10) does exist. The solution $B(z)$ which results cannot be created physically, for $B''(z)$ has to be positive everywhere (which means that the curve $B(z)$ must be convex towards the axis); however, it should be sufficient to produce the correct field in regions in which r_α is large, and to join it at either end to field-free regions as abruptly as possible.

In 1940, however, Recknagel pointed out that condition (b) above cannot be satisfied by such a lens: the image will always be virtual. The calculation is therefore invalid, and Scherzer's proposition remains completely valid. Recknagel also showed that the electrostatic lens free of aberration which would be obtained from (3) cannot form a real image either. Nevertheless, a field distribution $B(z)$ of the type derived by Glaser might be used to diminish the spherical aberration; successful attempts have been made to improve β-spectrographs in this way (Hubert, 1951; Mladjenović, 1953; Nadeau, 1951).

3. THE SEARCH FOR LENSES WITH LITTLE SPHERICAL ABERRATION

3.1 Weak Lenses with Minimum Aberration

(1). For a weak electrostatic einzel lens, we follow Scherzer (1936b) and write

$$\Phi(z) = \Phi_0\big(1 + \lambda\phi(z)\big) \qquad \text{with } \lambda\phi(z) \ll 1 \qquad (11)$$

$$\frac{1}{f} = \frac{3\lambda^2}{16} \int_{-\infty}^{+\infty} \phi'^2(z)\, dz, \qquad h = \frac{3\lambda^2}{16} \int_{-\infty}^{+\infty} z^2 \phi'^2\, dz. \qquad (12)$$

(f is the focal length and h the abscissa of the principal plane.)

The radius of the aberration patch in the Gaussian image plane, abscissa b, is then

$$\rho = -\frac{5\lambda^2}{64} b r_B^2 \int_{-\infty}^{+\infty} \phi''^2\, dz$$

in which r_B denotes the radius of the aperture within the lens. ρ has a minimum for

$$\Phi(z) = \Phi_0\big(1 + A e^{-Bz^2}\big). \qquad (13)$$

The values of A and B are chosen such that f and h have the desired values:

$$A = \pm 4\left(\frac{8h}{27\pi f}\right)^{\frac{1}{4}}, \qquad B = \frac{3}{4hf}.$$

We then find

$$\rho_{min} = \frac{15}{16} \cdot \frac{r_B^3 b}{f^2 h} \tag{14}$$

and it can be seen that ρ decreases if h increases, and hence, if the lens is made more convergent. The lens with least aberration ought to be a strong lens, but the method of calculation is not applicable to this case.

The equation of the equipotential surfaces far from the axis can be calculated, but unfortunately the electrodes cannot in practice be given the corresponding shapes.

(2). Glaser (1938) has calculated the distribution $B(z)$ which would give a weak magnetic lens with the least aberration; here, we have simply

$$\rho = \alpha^3 \frac{e}{16 m_0 \Phi_0} \int_{z_0}^{z_i} \left(\frac{e}{3 m_0 \Phi_0} B^4(z) + B'^2(z) \right) dz. \tag{15}$$

Again, the calculated field distribution cannot be produced in practice.

(3). A few years later, Plass (1942) attempted to discover what shape the electrodes would have to be given in order to produce the potential distribution derived by Scherzer, $\Phi(z) = \Phi_0(1 + Ae^{-Bz^2})$, in a reasonably simple way. He calculated the potential distribution near the axis using the classical series expansion

$$\Phi(r, z) = \Phi(z) - \frac{r^2}{4} \Phi''(z) + \frac{r^4}{64} \Phi''''(z) - \cdots$$

limited to terms in r^{12}; the electrodes which would correspond to three of the equipotential surfaces obtained are shown in Fig. 2. In the special case,

$$\Phi(z) = \Phi_0 \left(1 - 0.5 e^{-z^2/2} \right) \tag{16}$$

(z measured in cm), f is found to be 12.4 cm for a lens about 12 cm thick; the lens is certainly weak. For parallel incident rays, we find $\rho_1 = 0.625 u^3$ in the image focal plane, where u denotes the distance of the extreme rays from the axis at $z = 0$. The corresponding aberration constant is then $C_3 \simeq 145$ cm, or $C_s \simeq 12$.

Plass also gives the results of his calculations on an immersion lens, in which $\Phi(z) = \Phi_0(1 - \frac{1}{2}\tanh z)$, $f_1 = 18.8$ cm and $f_2 = 10.8$ cm. We now have $\rho_2 = 0.334 u^3$. An intuitive argument, originating in classical optics where the aberration of a centered system is reduced by combining glasses

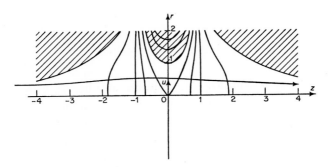

Figure 2 Plot of the equipotential lines corresponding to the axial potential function $\Phi(z) = \Phi_0(1 - 0.5e^{-z^2/2})$. Shaded areas indicate equipotentials suitable to be used in the construction of this lens.

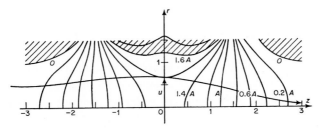

Figure 3 Plot of the equipotential lines in space which result from an axial potential function $\Phi(z) = \Phi_0(Ae^{-\left(\frac{z+0.85}{2}\right)^2} + Ae^{-\left(\frac{z-0.85}{2}\right)^2})$.

with different refractive indices, subsequently led Plass to examine a distribution of the form

$$\Phi(z) = \Phi_0\left(A_1e^{-\left(\frac{z+z_0}{2}\right)^2} + A_2e^{-\left(\frac{z-z_0}{2}\right)^2}\right) \tag{17}$$

acting on a suggestion of L. Marton. In the special case for which $z_0 = 0.85$ cm, the aberration passes through a minimum, and we find

$$\rho_3 = 0.334u^3 \cong 0.4\rho_1. \tag{18}$$

We can deduce that $C_s \simeq 6$.

The corresponding equipotentials, which show how such a lens can be designed, are depicted in Fig. 3. The lens is again weak, however.

(4). Recently, P. Lapostolle, working at CERN, has attempted to build an electrode system which will display very little aberration; the principle is very different however.

Within the lens, the outermost trajectories are far from the axis; the general equation of motion in a rotationally symmetrical system must therefore be used:

$$\frac{r''}{1+r'^2} + \frac{1}{2\Phi(r,z)} \cdot \frac{\partial\Phi(r,z)}{\partial z} \cdot r' - \frac{1}{2\Phi(r,z)}\frac{\partial\Phi(r,z)}{\partial r} = 0 \qquad (19)$$

in which

$$\Phi(r,z) = \Phi(z) - \frac{r^2}{4}\Phi''(z) + \frac{r^4}{64}\Phi''''(z). \qquad (20)$$

On the axis, the function $\Phi(z)$ can be written as a power series within the central zone of the lens (where the rays are furthest from the axis):

$$\Phi(z) = \Phi_0 + az^2 + bz^4 + \ldots$$

and hence

$$\Phi(r,z) = \Phi_0 + az^2 - \frac{a}{2}r^2 + bz^4 - 3br^2z^2 + \frac{3}{8}br^4. \qquad (21)$$

The term in r'^2 is neglected, as its contribution is of the order of 5%. The task of solving the general equation can then be simplified by seeking the type of potential distribution for which

$$\frac{1}{\Phi(r,z)}\frac{\partial\Phi(r,z)}{\partial z} \qquad \text{is independent of } r \qquad (a)$$

$$\frac{1}{\Phi(r,z)} \cdot \frac{\partial\Phi(r,z)}{\partial r} \qquad \text{is proportional to } r. \qquad (b)$$

Condition (a) is less important, since this term is multiplied by r' and is hence always small in the region where r is greatest. Condition (b) gives

$$-\frac{1}{\Phi(r,z)}\frac{\partial\Phi(r,z)}{\partial r} = \frac{a}{\Phi_0}r\frac{1+\frac{b}{a}\left(6z^2 - \frac{3}{2}r^2\right)}{1+\frac{a}{\Phi_0}\left(z^2 - \frac{1}{2}r^2\right)} = Kr. \qquad (22)$$

This condition cannot be satisfied for all the trajectories, but for a mean trajectory, it can be fulfilled; this leads one to hope that a satisfactory correction can be achieved for all the trajectories. If, under these conditions,

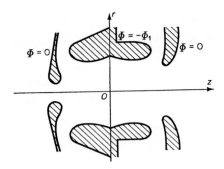

Figure 4 Weak aberration lens L_0 (Lapostolle, 1959).

we put

$$k = \frac{1}{\Phi_0} \frac{a^2}{b}, \qquad u = \frac{\Phi(z)}{\Phi_0}; \qquad x = \left(\frac{b}{a}\right) \cdot z, \qquad y = \left(\frac{b}{a}\right)^{\frac{1}{2}} r$$

the potential is given by

$$u = 1 + k\left(x^2 - \frac{y^2}{2} + x^4 - 3x^2 y^2 + \frac{3}{8} y^4\right)$$

from which the equation for the equipotential surfaces can be extracted.

Fig. 4 illustrates the electrode shape which would provide the requisite distribution in the central zone. We shall designate this lens L_0. The resemblance between the shape of the central electrode and that of Plass (Fig. 3) is to be noted.

(5). *Hyperbolic Lenses.* In conclusion, we should mention a study by Rüdenberg (1948) in which he drew attention to the properties of hyperbolic lenses. These are ideal symmetrical einzel lenses, in which potential $\Phi(r, z)$ is of the form

$$\Phi(r, z) = A\left(z^2 - \frac{r^2}{2}\right) + B. \tag{23}$$

The equipotential surfaces are hyperboloids of revolution (Fig. 5), and it is a relatively straightforward task to create such a potential distribution within a closed space. Inside such a lens, the radial field, E_r, is everywhere proportional to the distance from the axis; the spherical aberration is theoretically zero for rays incident parallel to the axis, and very small when the object is at a finite distance from the lens. In a real lens, however, the outer electrodes have openings in them, and it is difficult to calculate the effect of

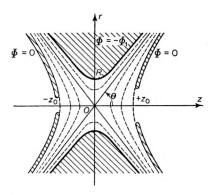

Figure 5 Plot of equipotential lines in an electron lens of hyperbolic field structure—($\theta = \tan^{-1}\sqrt{2}$).

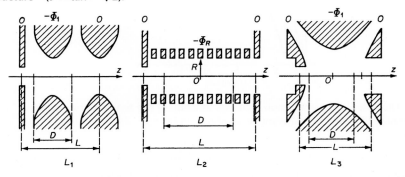

Figure 6 Three different lenses with approximately hyperbolic field structure in the central region. The approximation is good over a length D. (Left) Lens L_1: $D/L \simeq 0.5$; (Center) lens L_2: $D/L \simeq 0.6$ (the potential function $\Phi_R = \Phi_1\left(\dfrac{z^2 - (\frac{L}{2})^2}{(\frac{L}{2})^2}\right)$ is created over cylinder of radius R by means of several circular metallic rings); (Right) lens L_3: $D/L \sim 0.75$. The electrodes are hyperboloids.

the perturbed terminal regions. In a recent experimental study, using very big lenses and a special optical bench (Septier, 1957), we have measured the aberration constants of lenses of this type; the distribution $\Phi(r, z)$ is created in an approximate fashion over a wider and wider region, using electrodes of different shapes (Fig. 6), and we have compared the results with those given by the lens L_0 which has little aberration (Fig. 4).

It will be seen from Fig. 7, which summarizes our results (Septier, 1960), that the diameter of the aberration patch in the Gaussian image plane varies as α^3 in most of the lenses for values of α up to about 10°, and that terms of order higher than α^5 appear only at very large angles. The CERN lens L_0 is

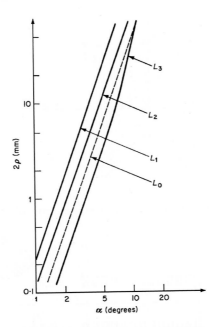

Figure 7 Diameter of the aberration spot as a function of the semi-aperture angle α, corresponding to L_0 and the different lenses shown in Fig. 6.

seen to be better than the cruder hyperbolic lenses (L_1, L_2) but not as good as a more carefully constructed hyperbolic lens (L_3), in which the influence of the end effects has been reduced. For these various lenses, we find

$$
\begin{array}{ccccc}
 & L_0 & L_1 & L_2 & L_3 \\
C_s & 12.5 & 50 & 25 & 8
\end{array}
$$

However, we must point out that once again the lenses are all comparatively weak, with their foci outside. If the electrode separation is reduced to the uttermost—about 2 mm for a potential of 50 kV on the central electrode—the minimum focal length of all these lenses is in the neighborhood of 10 mm.

3.2 Work on Strong Lenses

Whilst the ideal weak lens was being investigated, numerous other studies were being devoted to the determination of the optimum working conditions of extant electrostatic and magnetic lenses; in particular the lenses used as objectives in electron microscopes received much attention. It is, in fact, within the objective lens that the beams emerging from different points on the object have the largest angular aperture; in consequence, the theoretical resolving power of microscopes, δ_{th}, which is calculated by combining the aperture aberration and the diffraction aberration, depends upon the spherical aberration coefficient C_3 of this lens alone:

$$\delta_{th} = K(C_3)^{1/4} \cdot \lambda^{3/4}. \tag{24}$$

(K is a constant, and λ represents the wavelength of the particles: $\lambda = 12.2 \, V^{-\frac{1}{2}}$ when λ is measured in Å and V in volts.)

It is clear that C_3 must be considerably improved if δ_{th} is to be appreciably diminished. Since $C_3 = C_s f$, two means are available: either f or C_s may be reduced.

It is tempting to try to make f tend towards zero, by scaling down all the dimensions of the lens in question, while holding the applied potential or magnetic field constant. Various constraints limit such a procedure, however. In an electrostatic lens, the vacuum will break down if the inter-electrode distances are too small. With a magnetic lens, the convergence depends upon the maximum value, B_0, of the induction on the axis, and if the diameter, D, of the opening through which the beam passes between the poles is fixed, B_0 will decrease when the gap S in the lens is reduced (even if the excitation current of the lens is increased indefinitely); this is a consequence of pole-piece saturation, which limits the maximum induction in the gap to about 23 kG. Furthermore, this gap has to be large enough for the specimen-holder to be inserted. Finally, in both types of lens, a rapid increase in the fifth order aberration terms would ensue if such a reduction of the transverse dimensions were not accompanied by a similar reduction in the beam diameter.

We therefore seek lenses which have very short focal lengths, and at the same time, the smallest possible values of C_s.

(1). *Magnetic Lenses*. The properties of these lenses can easily be calculated, with the aid of approximate mathematical models; in particular, the simple "bell-shaped" model due to Glaser,

$$B(z) = \frac{B_0}{1 + (z_0/d)^2}$$

yields the aberration coefficient directly: Glaser (1941), Dosse (1941), and Lenz (1950, 1951) employ this method. As the excitation is increased, the

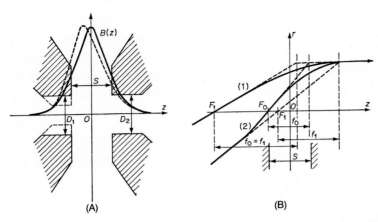

Figure 8 Strong magnetic lens. (A) Schematic cross-section of the lens, and shape of the axial magnetic field function $B(z)$ in two different cases: $D_1 = D_2$ (full lines) and $D_1 < D_2$ (dotted lines); (B) trajectories in the lens, the emergent ray being parallel to the axis Oz; F_1 and f_1 are the asymptotic elements, F_0 and f_0 the "immersed" ones. Ray (1) corresponds to a weak excitation, ray (2) to a strong excitation of the lens.

aberration at first falls rapidly and then passes through a very broad minimum. If d characterizes the half-width of the bellshaped curve, $B(z)$, it is found that in theory,

$$C_{3\,min} \simeq 0.25d \quad \text{with } f_0 \simeq d$$

or

$$C_s \simeq 0.25.$$

Here, f_0 is the immersion focal length (Fig. 8), and is itself a decreasing function of the excitation. When the lens is being used as an objective with a very high magnification, the object must be placed at the immersion focus F_0. In practice, the minimum is unattainable, since saturation restricts the value of B_0.

Liebmann and Grad (1951) have verified Glaser's predictions, using more realistic field distributions; the latter were obtained from a resistance network, for lens gaps of various different shapes, and the results are similar to those described above.

After making these measurements, Liebmann (1951) went on to determine the geometrical parameters of the lens which would have the smallest possible spherical aberration, in absolute magnitude; to this end, he investigated the effect of changing the scale of the lens (always supposing the accelerating potential Φ_0 and the maximum axial magnetic field B_0 to be constant). In these conditions, an absolute optimum value of the diameter D exists, which leads to minimum aberration, for each value of the ratio S/D. This minimum itself decreases as S/D is increased. The saturation of the iron is now a limitation, however, and it is not the maximum axial field B_0 which must be considered, but the magnitude of the magnetic induction between the pole-pieces of the lens, B_{iron} (B_{iron} is always greater than B_0). In these conditions, it is found to be better to use lenses for which S/D is greater than unity, and that for each excitation, there is an optimum value of S/D. The problem of determining the absolute minimum of C_s is a most complex one.

From the curves obtained by Liebmann, with $B_{iron} = 20\,\mathrm{kG}$, $V = 100\,\mathrm{kV}$, and $S/D = 2$, it is possible to obtain $C_s \simeq 0.5\,\mathrm{mm}$ with a corresponding focal length of the order of 1 mm. In this situation, however, the object has to be placed near the maximum of the field distribution $B(z)$, and this poses a new problem: the incident electron beam is powerfully focused onto the object by the part of the field which precedes it. We can then:

(i)   work with the object only slightly inside the field (this is the usual solution), although the spherical aberration coefficient then lies between 2 and 4 mm;

(ii) alternatively, we can reduce this focusing effect as far as possible by using a highly asymmetrical lens, as Liebmann (1951) suggested;

(iii) finally, we may on the contrary try to use it as the second condenser, with fixed convergence. Their attempts to find an objective with maximum convergence and minimum aberration led Ruska (1962) and Riecke (1962a, 1962b) to envisage this possibility (Fig. 9).

Riecke has shown that by adding a first condenser of variable focal length, an aperture, and a device with which the convergence of this first condenser can be adjusted with precision, such an objective can be used, and the practical resolving power of a magnetic microscope can be reduced to about 4 or 5 Å (Ruska, 1964, 1965) with an accelerating potential of 100 kV ($\delta_{th} \simeq 1.7\,\text{Å}$).

An examination of asymmetrical lenses (Dosse, 1941) reveals that when the object is only just inside the field, it is better to use a lens in which the axial variation of B is slower on the object side, for strong excitations; for the excitations normally used in objective lenses, however, all the lenses are virtually equivalent.

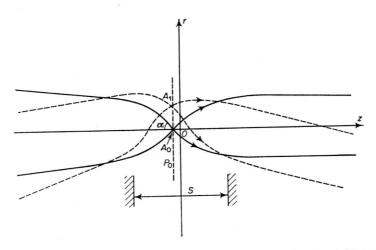

Figure 9 Trajectories in the very strong lens of Riecke and Ruska. P_0 = object plane, S = distance between poles.

Recent work by Durandeau and Fert (1957) and Dugas, Durandeau, and Fert (1961) has shown that the curves giving the spherical aberration of all the usual kinds of lenses can be reduced to a single curve, if a suitable system of reduced co-ordinates is employed; moreover, the same is true of the first order optical properties.

If the spherical aberration of magnetic lenses is to be still further reduced, far more intense fields will have to be made available, created by very small coils free of iron. Until recently, this solution seemed rather far-fetched, although Gianola (1950a) had already foreseen it; it would entail using pulsed operation and copper windings, which would scarcely be compatible with good resolution. With the discovery of hard superconductors, however, with critical fields of the order of 100–200 kG, it seems not unreasonable to foresee the construction of microlenses giving $B_0 = 100$ kG along a distance of several millimeters, disregarding the major technical difficulties; this would be of especial interest to the designers of very high voltage microscopes.

(2). *Electrostatic Lenses.* A general analysis, based upon a simple mathematical model, has been made by Glaser and Schiske (1954). Many articles contain values of the spherical aberration of individual lenses (Gobrecht, 1956; Gundert, 1939; Lippert & Pohlit, 1953; Mahl & Recknagel, 1944; Ramberg, 1942).

When the convergence increases, the aberration falls, passes through a minimum at the same time as f, and subsequently increases very rapidly with f. Experimental work by Heise (1949) has demonstrated this behavior very' clearly, and shows that the "constant" C_s also varies with the focal length f : C_s and f pass through their minima simultaneously. Liebmann (1949) has compared the properties of several symmetrical lenses, in a search for the types of lens which have minimum C_s. He showed that the aberration is smallest when the fringe fields at the ends of the lens are reduced to the utmost by providing the outer electrodes with grids. (Such lenses, cannot, however, be used when high resolution is required, since the openings in the grids create local aberrations.) Furthermore, the outer trajectories must not pass too close to the electrodes, and the initial divergence of the beam must be as small as possible to prevent fifth order aberration from having any effect.

Seeliger (1948, 1949) has also studied a series of einzel lenses of widely differing forms.

From all these investigations, it emerges that the best symmetrical electrostatic lenses have aberration coefficients some four or five times greater than those of magnetic lenses with external foci and the same convergence. For use as an objective (or as the last lens of a microanalyzer) the focus must be external, and the minimum focal length of such lenses is of the order of four to five millimeters. Since the minimum value of C_s is about 10, the coefficient C_3 of this type of lens will be

$$C_{3\,\mathrm{min}} \simeq 40 \text{ mm}$$

under the most favorable conditions.

3.3 Combinations of Lenses: Asymmetrical Lenses

(1). *Superposition of Electrostatic and Magnetic Lenses.* In 1937, Rebsch and Schneider announced the result of the first calculation to be made upon a complex lens, formed by superimposing an electrostatic field $\Phi(z)$ and a magnetic field $B(z)$. If G is a figure of merit, inversely proportional to the aberration ρ, f_{el} and f_{mag} denote the focal lengths of the isolated electric and magnetic lenses respectively, l_{el} and l_{mag} are factors proportional to the "length" of the distributions $\Phi(z)$ and $B(z)$ along the axis, and η is a

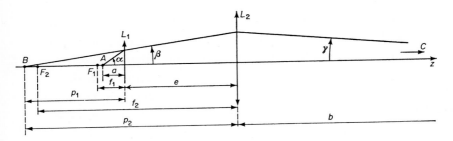

Figure 10 High magnification doublet.

dimensionless factor, then for the combined lens it is found that

$$G_{\text{total}} = G_e \frac{\left(1 + \eta^2 \dfrac{f_{\text{el}}}{f_{\text{mag}}}\right)^2}{\left(1 + \eta^2 \dfrac{\rho_{\text{el}}}{\rho_{\text{mag}}}\right)\left(1 + \eta^2 \dfrac{f_{\text{el}}}{f_{\text{mag}}} \dfrac{l_{\text{mag}}^2}{l_{\text{el}}^2}\right)}. \tag{25}$$

Further calculations show that

$$G_{\text{total}} < G_{\text{el}} \quad \text{and} \quad G_{\text{total}} < G_{\text{mag}}$$

are always true, irrespective of the values of the two quantities on the right-hand sides of the inequalities. The spherical aberration of a lens of this type will therefore always be greater than that of the electric or magnetic lens in isolation.

(2). *The Doublet.* Rebsch (1938) and Marton (1939) drew attention to a genuinely possible means of diminishing the aberration of centered systems consisting of two lenses, and L_1 and L_2. The focal lengths, f_1 and f_2 are such that L_1 produces a virtual image, B, of the object, A, while L_2 provides the final real image, C; f_1 is made to tend towards zero, or at least, is made as small as is compatible with the properties of the ferromagnetic material in the magnetic case, or with the avoidance of electrical breakdown in the electrostatic situation. The principle is as follows (Fig. 10): in the image plane C, the aberration is given by

$$\rho = C_s f \alpha^3 \leqslant \left[C_1 f_1 + C_2 f_2 \left(\frac{a}{p_1}\right)^4 \right] \alpha^3 \tag{26a}$$

$$C_s f \leqslant C_1 f_1 + C_2 f_2 \left(\frac{a}{p_1}\right)^4. \tag{26b}$$

The contribution from L_2 becomes negligible if $(a/p_1) \ll 1$, and this is so if $a \simeq f_1$. By reducing f_1, we can decrease $C_s f$ and hence the spherical aberration of the doublet, always keeping the magnification $M = \alpha/\gamma$ constant.

Marton and Bol (1947) have calculated the spherical aberration of such a doublet; each lens is represented by a Glaser bell-shaped function $B(z)$, and the two functions $B_1(z)$ and $B_2(z)$ do not interact, which entails leaving quite a wide interstice between them. Under these conditions, the convergence is always low, and the improvement in $(C_s f)$ does not exceed 20%. Marton stresses in conclusion the need to bring the lenses very close to one another, which renders any general calculation impossible: the distribution $B_3(z)$ which results from the superposition of $B_1(z)$ and $B_2(z)$ will be known only imperfectly, and cannot be assimilated into a simple analytic function.

(3). *Asymmetric Electrostatic Lenses.* If the axial potential distributions, $\Phi_1(z)$ and $\Phi_2(z)$, of two electrostatic lenses, L_1 and L_2, are brought sufficiently close together, an asymmetrical curve, $\Phi(z)$, results, which has its greatest slope on the L_1 side, the object side. Hanszen (1958a) has demonstrated experimentally, by means of a highly asymmetrical lens, that it is better to place the object on the side where $\Phi'(z)$ reaches its greatest value (if high magnification is required), and has set out a procedure for determining the two lenses L_1 and L_2 which would comprise the doublet equivalent to a given asymmetric lens (Hanszen, 1958b). In fact, this is a rediscovery of a result well-known in classical optics: at high magnifications, the spherical aberration is least when the face of the lens with the greatest curvature is turned towards the object, which is thus closer to the region of most rapid convergence.

This remark led us to investigate asymmetrical lenses experimentally, with a view to lowering C_s as far as possible (Septier & Ruytoor, 1959).

Fig. 11 shows diagrammatically a cross-section of one of the best lenses obtained, and Fig. 12 illustrates the large influence of a small variation of the internal diameter of the central electrode: lens L'_4 differs from L_4 only in that this diameter has been reduced by 16%. It was found that

$$C_s = 4 \quad \text{for } L_4$$
$$C_s = 2 \quad \text{for } L'_4$$

but at large angles, an important supplementary term in α^5 intervenes. A lens such as this could be employed with advantage as the final stage of an electron probe system, for the focus is external. The minimum focal

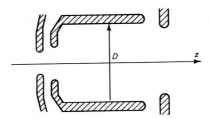

Figure 11 Cross-section of the asymmetric electrostatic unipotential lens L_4.

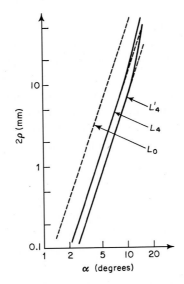

Figure 12 Transverse aberration of the asymmetric electrostatic lenses L_4 and L'_4 (the internal diameter D of L'_4 is 16% smaller than that of L_4).

length is of the order of 10 mm, so that $C_{3\,min} \simeq 20$ mm, a twofold gain with respect to the best symmetrical lenses.

Seeliger (1948) had already mentioned that one of the asymmetrical lenses he had studied was superior to symmetrical lenses; he obtained $C_s \simeq 4$. This superiority arises from the fact that the optical center of the lens can be brought closer to the object (in our own lenses, it lay practically in the median plane of the entry diaphragm of the central electrode) and that the maximum diameter $2R_m$ of the beam inside the lens is smaller than it would be in a symmetrical lens having the same convergence—this reduces the overall aberration. The ratio of the values of R_m in L_0 and L'_4 corresponding to the same angular aperture α and the same convergence is such that

$$\frac{[(R_m)L_0]^3}{[(R_m)L_4']^3} \simeq 5;$$

to all intents and purposes, this corresponds to the ratio of the aberration constants.

The form of the potential distribution $\Phi(z)$ which will yield minimum aberration for strong electrostatic lenses has been calculated by Tretner (1959). This distribution could be produced in an approximate fashion by means of a very asymmetrical lens, in which the maximum of the field $E_z = -\Phi'(z)$ would be very close to the object. This confirms the preceding results very satisfactorily.

(4). *Conclusion*. The conclusion to be drawn from this discussion is that despite all the research which has been devoted to them, and which seems to have exhausted all the possibilities of improvement, electrostatic lenses are markedly inferior to their magnetic counterparts; operating with an external focus,

$$(C_{3\,min})_{electrostat} = 20 \text{ mm} \simeq 5(C_{s\,min})_{mag}$$

and we have seen that for a magnetic objective with a deeply immersed focus of the type employed by Ruska, it is possible to reach

$$C_{3\,min} \simeq 0.5 \text{ mm}.$$

To obtain a major improvement, therefore, we must consider methods of achieving complete correction of the aperture aberration.

4. ATTEMPTS TO CORRECT THE APERTURE ABERRATION

In Section 2.3, the conditions for the validity of formula (9) were set out; this formula was established on the basis of *geometrical* optics. Scherzer himself has summarized the principal ways of overcoming the difficulty:
(a) by using high frequency potentials;
(b) by introducing space charge;
(c) by employing rotationally symmetrical optical systems in which the function Φ'/Φ has singularities;
(d) by abandoning rotational symmetry.

Some of these procedures are more promising than others, and their practical worth increases from (a) to (d). The techniques (a) and (b) have the same origin: $\Phi(z)$ will no longer satisfy Laplace's equation, but Poisson's equation, or the equation of propagation of high frequency waves.

We shall now review these various techniques in order of increasing promise; we shall dwell in particular upon the correction methods which involve either local or wholesale departure from rotational symmetry.

Finally, we shall mention a last possibility, which is a direct application of the methods of *physical* optics to electron optics. The aberration patch which surrounds a point imago may be regarded as a diffraction disc, enlarged by the aperture aberration of the objective lens. Having suppressed the aperture aberration of the lens, therefore, we should recover the pure diffraction disc, which could, furthermore, be reduced by increasing the useful aperture of the objective. Recent attempts have shown that in theory, the spherical aberration could be virtually annulled by using a zone plate, similar to the Fresnel zone plates of classical optics.

4.1 High Frequency Lenses

Electrons which leave a point source S at the same moment and with the same speed will have a range of transverse velocities, and will therefore arrive at different times at the object plane of an electrostatic lens, L. If a very high frequency potential Φ_f is superimposed upon the steady excitation potential of the lens, and if the phase of Φ_f is suitably chosen, the electrons which travel farther from the axis, and which arrive later than those remaining close to the axis, will experience a weaker converging force, and will hence be less deflected. Another possibility is to increase the velocity of the outer electrons, without changing that of the paraxial ones, by using a modulating twin-grid gap, perpendicular to the beam and supplied with the h.f. voltage Φ_f (Kompfner, 1941, see Fig. 13(A)).

Since the time differences involved are extremely small, very high frequencies will have to be employed (centimeter waves). Furthermore, the incident beam must be cut up into very short pulses, with the aid of a pulsed gun, for example Zworykin, Morton, Ramberg, Hillier, and Vance (1945); their length τ must be very short in comparison with the period, T, of the applied u.h.f. potential: $\tau \ll T$. A variable phase-shifter will enable the phase difference between the trigger voltage of the gun and the potential applied to the lens to be adjusted. Since the requisite frequency lies between 3000 and 10,000 Mc/s, the difficulty of the problem of obtaining the short pulses for the gun supply is immediately obvious ($\tau \ll 10^{-10}$ s). The lens could be built in the way described by Scherzer (1947), who studied the operating conditions of weak lenses of this kind in detail (see

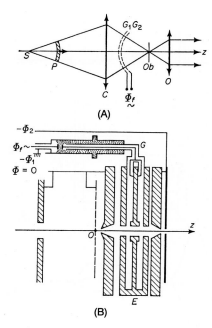

Figure 13 High frequency lenses. (A) Kompfner's system. S = pulsed electron source; P = infinitesimally thin packet of monokinetic electrons traveling towards the lenses. C = condenser lens. Ob = object; O = objective lens to be corrected. Owing to the different trajectory lengths, marginal and paraxial electrons cross the gap formed by the grids G_1 and G_2 at different times. (B) Scherzer's lens—the h.f. voltage is applied through a wave-guide to the middle part of the central electrode, the whole forming a resonant cavity.

Fig. 13(B)). For $f_0 = 10,000$ Mc/s and a u.h.f. potential of 500 V, a corrected objective with short focal length ($f \simeq 5$ mm) could be constructed. No experimental check has as yet been attempted, however, despite recent progress in electronics.

4.2 Electrostatic Charge

(1). *Space Charge.* If a charge distribution of density $\rho = \rho(r, z)$ is placed around the axis of the optical system, the electron trajectories will be given by the equation

$$\Phi(z)r'' + \frac{1}{2}\Phi'(z)r' + \frac{1}{4}\Phi''(z)r + \frac{1}{4\varepsilon_0}\rho(r, z)r + P(z, r^3) = 0 \qquad (27a)$$

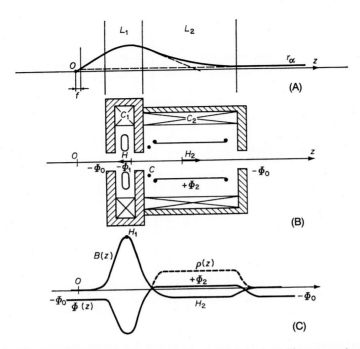

Figure 14 Space charge lens proposed by Ash and Gabor. (A) Electron trajectories in the lens. L_1 = converging magneto-electric lens; L_2 = diverging space-charge lens. (B) Schematic cross-section of the lens. The electron cloud is axially confined by the axial magnetic field between an emissive cathode C and the central tubular anode. (C) Curves giving distribution of potential and magnetic field along the axis Oz.

in which $P(z, r^3)$ denotes the terms responsible for the aperture aberration. If the distribution $\rho(r, z)$ is chosen correctly, it is possible to cancel the aberration of the electrostatic lens in question; ρ must increase from the axis towards the periphery: $\rho \simeq \rho_0 + Ar^2$. Gabor was the first to point out this possibility (1945a, 1945b); he suggested using an arrangement consisting of a mixed magnetic and electrostatic lens (Fig. 14), behind which would be placed a device to form a stable dense cloud of electrons; the latter would behave as a strong divergent lens, with the result that

(i)   the focal length of the whole system would be a few microns (Fig. 14(A)); this would already reduce $C_3 = C_s f$;

(ii) the combined system would be corrected for spherical aberration ($C_s \simeq 0$).

Ash and Gabor (1955) produced a high density electron cloud with axial symmetry experimentally, and measured the broadening of the spot

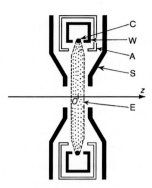

Figure 15 Space charge lens of Haufe (schematic). C = annular cathode, W = Wehnelt electrode, A = anode, S = shield, E = electron cloud.

due to the scattering of the incident electrons by those in the cloud. The effect of the fluctuating nature of the medium, the "frosted-glass" effect, resulted in an aberration patch, the approximate diameter of which is given by Scherzer (1947):

$$d \simeq M \left(\frac{e}{4\varepsilon_0 \Phi_0} \int_{z_0}^{z_i} \frac{\rho(r, z)}{\Phi(z)} r_\alpha^2 \, \mathrm{d}z \right)^{\frac{1}{2}}. \tag{27b}$$

In the system of Ash and Gabor, the presence of positive ions leads to fluctuations in the density distribution $\rho(r, z)$. An attempt to achieve genuine correction has recently been made by Haufe (1958), who used a special short lens which produced a thin layer of charge perpendicular to the axis, rotationally symmetric and perfectly stable (Fig. 15). By examining the shadow of a grid, Haufe showed that it was possible to correct an ordinary lens placed beyond his device, in the central zone at least: the variation of ρ with r was unsatisfactory, and complete correction for every point was not possible. This partial correction enabled Haufe to reduce the size of the spot produced by the lens by a factor two; despite a magnification of some 2500 times, Haufe was not able to discern any "frosted-glass" effect due to the space charge.

No experimental work on a microscope or micro-probe device has as yet been attempted.

(2). *Induced Charges on Conductors.* (i) We could attempt to obtain a similar correction by placing an insulated metal foil on the axis of the lens, close to a metal diaphragm held at a positive potential. Negative charge accumulates over the foil, and the charge density is greater farther from the axis; it can thus be used to correct the spherical aberration of a lens placed in its vicinity. The beam is liable to be affected by its passage through the foil, however, even if the latter is very thin, and additional aberration due to scattering may result. No experimental work on this method has as yet been attempted, although it is now possible to produce extremely thin (50–100 Å) beryllium foils (Hast, 1948).

(ii) One other possible technique has been studied, theoretically at least (Hubert, 1949): the beam of incident electrons is passed through an insulated metal tube, in order to exploit the attraction between the electrons and their image charges. This method of correction would be feasible only with very slow electrons; otherwise, the diameter of the tube would have to be a few microns only, over a length of the order of a centimeter, and this effectively destroys any practical value the method might possess.

4.3 Discontinuities in the Function $\Phi'(z)/\Phi(z)$

There are two cases to consider: either $\Phi'(z)$ has a large enough discontinuity, or there are regions in which $\Phi(z)$ vanishes. The former corresponds to grid (or gauze) lenses (or to a conducting foil connected to a potential source); the latter corresponds to electrostatic mirrors, and coaxial lenses.

(1). *Grid (or gauze) lenses.* In a symmetrical einzel lens, with a central grid, the appearance of the potential distribution $\Phi(z)$ is shown in Fig. 16; if the potential, Φ_G, of the central electrode has a retarding effect upon the incident electrons, the lens will be diverging, whereas if the electrons are accelerated, it will be converging. Assuming an approximate expression for the potential distribution, $\Phi(z)$, Scherzer (1947) has shown that the aberration of a weak divergent lens of this kind could be overcorrected, if the thickness, e, of the central electrode and the potential Φ_G, were chosen judiciously; such a lens could then be used to correct a standard round lens.

Bernard (1952, 1953) has made a very thorough examination of grid lenses, first assuming the grid to behave like a thin metal membrane. He has shown that the general expression for C_3 calculated by Scherzer is rendered invalid by the discontinuity in $\Phi'(z)$ at the grid, and he has calculated the correct expression for the coefficient, which now contains supplementary negative terms. It is thus possible to cancel, or even reverse the sign of C_3 by altering the potential on the grid. This is possible only when the lens is divergent, however; a grid lens free of aberration cannot be used on its own, but only as a correcting element.

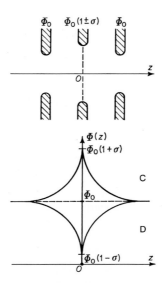

Figure 16 Unipotential grid-lens (Bernard): cross-section, and axial potential function $\Phi(z)$. The lens may be convergent (curve C) or divergent (curve D).

Later, Bernard calculated the aberration which would be introduced by the presence of openings in the grid. In the image plane, this aberration is, disappointingly, of the same order of magnitude as the diameter of the holes (a few microns).

Another corrector, using two parallel thin foils held at different potentials, has been suggested by Gianola (1950b). Again, it does not seem possible to exploit the theoretical correction possibilities, since the foils or grids introduce larger defects than those they are intended to eliminate.

(2). *Combination of Lens and Mirror.* An examination of the spherical aberration of an electrostatic mirror reveals that the sign of this aberration is the reverse of the sign of the corresponding aberration of a lens. The centered system of Fig. 17, for example, can be shown to be perfectly corrected (Zworykin et al., 1945; Ramberg, 1942) if the dimensions are as shown, if the lens field is assumed to vanish to the right of the entry plane of the mirror, and if the spherical aberration of the lens is given by $\rho = -4d\theta^3$ (this implies that the spherical aberration constant C_s is of the order of unity, which can be achieved with a magnetic lens). The image (or object) lies in a region where the field is intense, however, and in practice, a system cannot be built to this design. It might be possible to separate

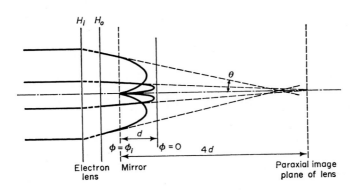

Figure 17 Combination of lens and concave mirror.

mirror and lens slightly, by using a weakly convergent mirror (external focus) with very high aberration; the combination would be only very weakly convergent, however, and a second lens would have to be used, through which the beam illuminating the object would also have to pass. The technical problems are so difficult that no practical solution has as yet been proposed. The simpler system, in which a concave mirror replaces the projective lens, quite a considerable distance behind a normal objective, cannot be adopted; the aberration correction produced by the mirror is now negligibly small, since the angular aperture of the beam which reaches it is so slight.

(3). *Coaxial Lenses*. If an electrode is placed along the axis of a symmetrical einzel lens (Fig. 18(A)), and held at a constant potential—that of the incident particles, for example—a divergent lens-element can be created, the focal length of which is a function of the off-axial distance. The means of supporting the axial electrode would have to lie outside the correcting element in field-free space, to prevent any disturbance of the symmetry within the useful region. The electron beam might be slightly inclined to the axis inside the lens to be corrected, so that it avoided the central electrode and its supports; thus one could employ a hollow conical beam having the same axis as the system itself, with angular aperture α_0 at the apex and angular extent $\Delta\alpha$ (see Fig. 18(B)).

Only a privileged annular zone of mean radius $\vec{r}_0$ can be corrected. The following conditions must then be satisfied in the neighborhood of $\vec{r}_0$:

$$\frac{\delta\alpha}{\delta r} = \frac{\delta^2\alpha}{\delta r^2} = \frac{\delta^3\alpha}{\delta r^3} = 0. \tag{28}$$

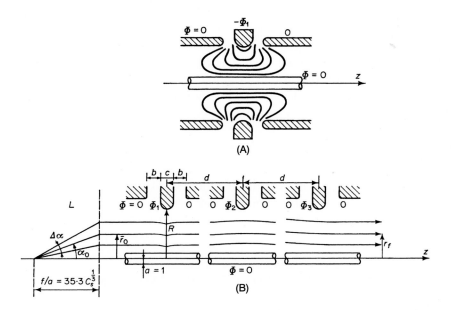

Figure 18 Coaxial lens. (A) Plot of equipotential lines in a unipotential lens with an axial electrode. (B) Zonally corrected lens system, using three coaxial lenses (Gabor). a = wire radius, $b = 0.66a$, $c = 1.34a$, $d = 8a$, $R = 3.66a$, $r_f = 1.97a$. Cathode at -30 potential units, $\Phi_1 = +0.40$, $\Phi_2 = -1$, $\Phi_3 = +0.5274$.

In coaxial lenses, the angular deviation of the rays varies as $1/r$, to the first approximation (whereas for the paraxial trajectories in an ordinary lens, it varies as r). To satisfy these conditions, a combination of three successive coaxial elements has to be placed beyond the lens to be corrected. The system suggested by Gabor (1946) and illustrated in Fig. 18(B) has been studied in detail by Dungey and Hull (1947).

Calculation shows that the aberration is well–corrected over the zone

$$1.845a < r < 2.115a$$

(where a is the radius of the central conductor). The residual aberration patch will be 3 Å across, for a hollow beam completely filling this zone, provided that the mechanical precision of the numerous electrodes is high enough for no additional defects to appear. No experimental work on such a corrector has been attempted.

A simplified system with a single coaxial element has, on the contrary, been recently studied in practice by Dupouy and Trinquier (1962); the element consists of a wire along the axis surrounded by an outer tube of radius r_1, and either an electrostatic or a magnetic version can be used (in the latter, a current flows along a wire and produces a field $H_0 \propto 1/r$). The spherical aberration of a round magnetic lens placed in front of the corrector can be compensated, but only over a much narrower zone than in the preceding case.

4.4 Departure from Rotational Symmetry

This was the first means of correction that Scherzer suggested (1947), and it immediately awakened very considerable interest. Much work has been devoted to this proposal, both theoretical and experimental, and it is for this reason that we shall study it in some detail, and accord it a separate section.

5. CORRECTION BY MEANS OF ASTIGMATIC SYSTEMS

5.1 The General Principle

The potential of a system with four planes of symmetry is obtained in the third-order approximation if we limit the expansion to terms of the fourth degree in r; if we employ (r, θ) co-ordinates, and choose the origin of θ to be the x-axis, which lies in one of the symmetry planes, we find:

$$\Phi(r, \theta, z) = \Phi - \frac{1}{4}\Phi''r^2 + \Phi_2 r^2 \cos 2\theta + \frac{1}{64}\Phi''''r^4 - \frac{1}{12}\Phi_2''r^4 \cos 2\theta$$
$$+ \Phi_4 r^4 \cos 4\theta \tag{29}$$

or in Cartesian co-ordinates:

$$\Phi(x, y, z) = \Phi_0 + \left(\Phi_2 - \frac{1}{4}\Phi''\right)x^2 - \left(\Phi_2 + \frac{1}{4}\Phi''\right)y^2$$
$$+ \frac{1}{64}\Phi''''(x^2 + y^2)^2 - \frac{1}{12}\Phi_2''(x^4 - y^4)$$
$$+ \Phi_4(x^4 - 6x^2y^2 + y^4) \tag{30}$$

so that

$$\Phi(x, y, z) = \phi_0(x, y, z) + \phi_2(x, y, z) + \phi_4(x, y, z) \tag{31}$$

where

$$\phi_0(x, y, z) = \Phi(0, 0, z) - \frac{1}{4}\Phi''(0, 0, z)(x^2 + y^2) + \frac{1}{64}\Phi''''(0, 0, z)(x^2 + y^2)^2$$

$$\phi_2(x, y, z) = \Phi_2(0, 0, z)(x^2 + y^2) - \frac{1}{12}\Phi_2''(0, 0, z)(x^4 - y^4)$$

$$\phi_4(x, y, z) = \Phi_4(0, 0, z)(x^4 - 6x^2y^2 + y^4). \tag{32}$$

Φ'' and Φ'''' denote derivatives of Φ with respect to z.

A system with four planes of symmetry can thus be regarded as the superposition of a round lens, the axial potential of which would be given by $\phi_0(x, y, z)$, a quadrupole lens, corresponding to $\phi_2(x, y, z)$, and an octopole lens, $\phi_4(x, y, z)$.

In a round lens, the radial force is proportional to r in the Gaussian approximation; further from the axis, however, where the term in r^3 has an effect, the rays are more strongly converged. It might be thought that a defocussing force proportional to r^3 could be introduced by means of the $\Phi_4(x, y, z)$ term, but it is a consequence of Laplace's equation that this term necessarily varies as $\cos 4\theta$; if, for example, the force has a defocussing effect in the x- and y-directions, it will focus the beam in the directions at $45°$ to these, so that if the aberration in the Ox, Oy directions is canceled, it will be doubled along the bisectors of the angle xOy.

5.2 Historical Survey of the Various Attempts at Correction

(1). *Scherzer's System.* Scherzer has shown (1947) that by using a combination of different systems containing both ϕ_2 and ϕ_4, a round lens can indeed be corrected. The distributions $\phi_2(z)$ and $\phi_4(z)$ are not superimposed onto $\phi_0(z)$, but are placed along the axis behind it. The principle is then as follows: the round lens is followed by a strongly convergent astigmatic system, giving two real line foci, in the directions Ox and Oy. In each of the two planes in which these focal lines are formed, a correction element (producing the ϕ_4 terms) is placed; the latter affects only rays which are far from the axis, and hence acts on the rays in one plane alone. A further astigmatic system renders the beam rotationally symmetrical again, and a final corrector is placed at the exit, at $45°$ to the other two, to correct any residual aberration along the bisectors of the angle xOy. By varying the relative magnitudes of the three corrector potentials, the spherical aberration can be wholly corrected.

Let x_a denote the trajectory lying in the plane xOz, for which $x_a = 0$ and $x_a' = 1$ in the object plane, and y_a the trajectory lying in the plane yOz, for which $y_a = 0$ and $y_a' = 1$ at the origin; an arbitrary trajectory which emerges from the point object A at an angle r_0' can then be described by

$$x = \alpha x_a \qquad y = \beta y_a$$

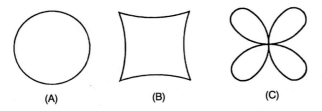

(A)   (B)   (C)

Figure 19 Elementary aperture aberration figures of a system with two planes of symmetry: (A) spherical aberration; (B) star; (C) rosette.

where α and β are the projections of r_0' on xOz and yOz respectively.

If we consider all the rays which emerge from A at the same slope, r_0', a rotationally symmetrical system will give a circular aberration patch in the image plane,

$$x_b = a\alpha(\alpha^2 + \beta^2)$$
$$y_b = a\beta(\alpha^2 + \beta^2)$$

of radius

$$\rho = a(\alpha^2 + \beta^2)^{3/2}.$$

A single coefficient, a, characterizes the aberration and in the Gaussian image plane, the aberration spot is a circle. When rotational symmetry is abandoned, we have

$$x_b = a\alpha^3 + b\alpha\beta^2$$
$$y_b = c\beta^3 + d\alpha^2\beta$$

(33)

with $b = d$, if the symmetry is altered in only two planes (Burfoot, 1954). The aberration is now characterized by three coefficients, a, b, and c, which must be annulled separately. The overall aberration pattern is now a more or less complex combination of three basic figures—circle, star, and rosette (Fig. 19).

If we consider a potential distribution consisting of ϕ_0 and ϕ_2, together with the correction term ϕ_4, the equations of motion are of the form:

$$\Phi x'' + \tfrac{1}{2}\Phi'x' - (\Phi_2 - \tfrac{1}{4}\Phi'')x = 2\Phi_4(x^3 - 3xy^2) = S_x$$
$$\Phi y'' + \tfrac{1}{2}\Phi'y' + (\Phi_2 + \tfrac{1}{4}\Phi'')y = 2\Phi_4(y^3 - 3x^2y) = S_y.$$

(34)

S_x and S_y are small perturbations, which are calculated using $x = \alpha x_a$ and $y = \beta y_a$. The influence of the corrector is given by

$$(x_b)_c = \frac{2M}{\Phi_a^{\frac{1}{2}}} \int_{z_a}^{z_b} \frac{\Phi_4}{\Phi^{\frac{1}{2}}} \left(3\alpha\beta^2 x_a^2 y_a^2 - \alpha^3 x_a^4 \right) dz$$

$$(y_b)_c = \frac{2M}{\Phi_a^{\frac{1}{2}}} \int_{z_a}^{z_b} \frac{\Phi_4}{\Phi^{\frac{1}{2}}} \left(3\alpha^2 \beta x_a^2 y_a^2 - \beta^3 y_a^4 \right) dz. \tag{35}$$

By adjusting Φ_4, we can arrange that

$$\begin{aligned} x_b + (x_b)_c &= 0 \\ y_b + (y_b)_c &= 0 \end{aligned} \tag{36}$$

in the following way: where $x_a = y_a \neq 0$, we place $\Phi_4 < 0$, and this enables us to cancel the terms in $\alpha\beta^2$ and $\alpha^2\beta$ (since $b = d$); in each of the two planes where $x_a = 0$ with $y_a \neq 0$ and $y_a = 0$ with $x_a \neq 0$, we place $\Phi_4 > 0$, and thus cancel the terms in β^3 and α^3 respectively, which have in fact been adversely affected by the first corrector.

The full calculation of the potentials which must be applied is extremely complicated. The value of $\Phi_4(z)$ is calculated from the trajectories which have been determined in first order from $\Phi_0(z)$ and $\Phi_z(z)$.

Rigorous calculation yields the following formula for the aberration patch which is finally obtained, after the beam has traversed the complex system:

$$\begin{aligned} x_b = {} & \frac{M\alpha^3}{16\Phi_b^{\frac{1}{2}}} \int_{z_a}^{z_b} \Phi^{\frac{1}{2}} \left[\frac{5}{4} \frac{\Phi''^2}{\Phi^2} - \frac{3}{2} \frac{\Phi'^2}{\Phi^2} \frac{x_a'^2}{x_a^2} + \frac{14}{3} \frac{\Phi'^3}{\Phi^3} \frac{x_a'}{x_a} + \frac{5}{24} \frac{\Phi'^4}{\Phi^4} + \frac{56}{3} \frac{\Phi}{\Phi^2} \right. \\ & \left. - \frac{8\Phi_2}{\Phi} \frac{x_a'^2}{x_a^2} - \frac{1}{2} \frac{\Phi_2 \Phi'^2}{\Phi^3} - 10 \frac{\Phi_2 \Phi''}{\Phi^2} - 24 \frac{\Phi_2 \Phi'}{\Phi^2} \frac{x_a'}{x_a} - 32 \frac{\Phi_4}{\Phi} \right] x_a^4 \, dz \\ & + \frac{M\alpha\beta^2}{16\Phi_b^{\frac{1}{2}}} \int_{z_a}^{z_b} \Phi^{\frac{1}{2}} \left[\frac{5}{4} \frac{\Phi''^2}{\Phi^2} + \frac{7}{4} \frac{\Phi'^2}{\Phi} \left(\frac{x_a'^2}{x_a^2} + \frac{y_a'^2}{y_a^2} \right) - 5 \frac{\Phi'^2}{\Phi^2} \frac{x_a' y_a'}{x_a y_a} \right. \\ & + \frac{7}{3} \frac{\Phi'^3}{\Phi^3} \left(\frac{x_a'}{x_a^2} + \frac{y_a'}{y_a} \right) + \frac{5}{24} \frac{\Phi'^4}{\Phi^4} - 8 \frac{\Phi_2^2}{\Phi^2} + 12 \frac{\Phi_2}{\Phi} \left(\frac{x_a'^2}{x_a^2} - \frac{y_a'^2}{y_a^2} \right) \\ & \left. + 4 \frac{\Phi_2 \Phi'}{\Phi^2} \left(\frac{y_a'}{y_a} - \frac{x_a'}{x_a} \right) + 96 \frac{\Phi_4}{\Phi} \right] x_a^2 y_a^2 \, dz \end{aligned} \tag{37}$$

with an identical expression for y_b, except that x_a and y_a, α and β, and Φ_2 and $-\Phi_2$ are interchanged; the second integral is unaffected by these

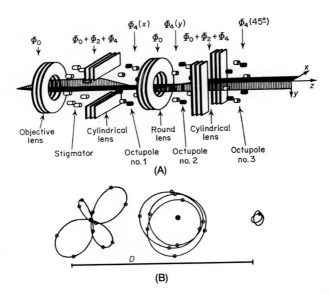

Figure 20 Seeliger's experimental system. (A) Schematic cross-section; (B) evolution of the aberration patch during the adjustment procedure. D indicates the diameter of the original aberration disc.

exchanges, which brings out very clearly the condition $b = d$ mentioned above.

An electrostatic version of this optical combination was assembled for the first time in 1949, and re-built in its definitive form between 1951 and 1953 by Seeliger (1951, 1953). Apart from the objective lens to be corrected, R_1 the system (see Fig. 20(A)) consists of a stigmator, S (a lens with several electrodes, the function of which is to correct the ellipticity astigmatism of R_1 and any residual astigmatism of the corrector), two cylindrical einzel lenses, with slits, L_x and L_y, a second round lens, R_2, and finally, three octopole correctors, O_x, O_y, and $O_{45°}$. The last of these is placed at the exit, at a point where the beam has already recovered its rotational symmetry.

The spherical aberration constant of R_1 was $C_3 = 92$ mm for $f_0 = 8.6$ mm; the overall aberration of R_1, L_x, L_y, and R_2 was such that $C_3 = 126$ mm (the focal lengths of the three lenses L_x, L_y, and R_2 were all equal to 15 mm).

Seeliger's experimental work entailed a most complicated sequence of successive adjustments of the various potentials applied to the electrodes. An extremely fine beam of electrons was directed into the objective at an angle of 2.6×10^{-2} radians; this beam could be made to describe a cone with its apex at the point object lying on the axis, and Seeliger could thus follow the evolution of the shape of the aberration patch (magnified by the projective lenses) as the adjustments proceeded. At the best adjustment, with the octopole correctors in action, the patch was reduced to 6% of its initial size (Fig. 20(B)).

Seeliger also measured the constant C_5 of the corrected system: $(C_5)_x = 21$ m, $(C_5)_y = 6$ m, and $(C_5)_{45°} = 13$ m, which would in theory allow a resolving power of 2 Å to be attained. He was unable to achieve a higher resolving power than that of the uncorrected device, but he did obtain this same resolution with an angular aperture three or four times as large.

These difficult and exacting experiments have the advantage of showing that the use of such a system would not give rise to additional aberrations, despite the presence of two violently astigmatic intermediate images. The experiments were revived by Möllenstedt in 1956, who used the same arrangement in an electrostatic microscope. By using an angular aperture of 2×10^{-2}, he was able to increase the resolving power of the instrument by a factor seven (at the same angular aperture); he was prevented from going beyond 30 Å, however, by vibrations and parasitic alternating fields. The experiment ought to be repeated on an instrument of a more suitable design.

(3). *Simplified Correctors.* The correction system of Seeliger entails setting up and adjusting seven additional lenses with high precision. Several attempts to simplify the system have been made, at least theoretically. We shall first mention the work of Burfoot (1953) and Whitmer (1956), who have investigated the possibility of creating the perturbations $\Phi_2(z)$ and $\Phi_4(z)$ *within the very lens to be corrected.* We shall then review the systems suggested by Archard (1954a, 1955, 1958) and Glaser (1955), before considering the most recent efforts; the latter are the results of the extensive experimental work which has been carried out in parallel by Deltrap at Cambridge, and Septier and Dhuicq in Paris.

(a) *The Burfoot Lens* (1953). With a view to reducing the number of adjustments to the minimum, Burfoot has calculated a design for an electrostatic lens with four electrodes of very special shape; such a lens produces the desired potential distribution, for which the third order aperture aberration is corrected. The fundamental $\Phi_0(z)$ appears at the first three electrodes, while $\Phi_2(z)$ is produced by the first, third, and fourth. Finally, $\Phi_4(z)$ is created at the central electrode. Fig. 21 shows the trajectories of the outer rays in the planes xOz and xOy. Unfortunately, the shapes of

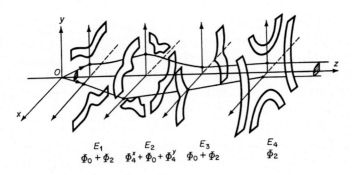

$$
\begin{array}{cccc}
E_1 & E_2 & E_3 & E_4 \\
\Phi_0 + \Phi_2 & \Phi_4^x + \Phi_0 + \Phi_4^y & \Phi_0 + \Phi_2 & \Phi_2
\end{array}
$$

Figure 21 Burfoot's lens (schematic).

the electrodes are so complicated, and the mechanical tolerances on these shapes and their alignment are so strict (considerably smaller than a micron), that to construct such a lens does not seem possible.

(b) *The Whitmer Lens* (1956). Whitmer has made a thorough but highly simplified examination of a perturbed lens of a similar type. However, each of his numerical calculations, which were based on perturbations $\Phi_2(z)$ and $\Phi_4(z)$ and could be produced experimentally, showed that the residual aberration patch would be asymmetrical, and larger than the aberration patch of the unperturbed lens. Neither in this nor in Burfoot's lens does the beam form line foci, whereas in Seeliger's system, the reverse is the case.

(c) *The Archard lenses* (1954a). Archard's first suggestion (system 1) was to replace Seeliger's cylindrical lenses by quadrupole lenses, together with an additional round lens (Fig. 22(A)), while retaining the same ray paths.

He next showed (1955) that the number of additional lenses could be reduced to four or five (systems 2 and 3, see Figs. 22(B) and (C)). These lenses are either purely quadrupole lenses (Φ_2) or eight-electrode lenses creating both Φ_2 and Φ_4.

Archard has studied in detail the optical properties of elements with four or eight electrodes, in the cases where the latter are spheres or thin plates (1954b).

More recently (Archard, 1958), he has suggested that the quadrupoles should also be used to produce the fundamental, $\Phi_0(z)$; the lens to be corrected is thus completely suppressed, being replaced by a self-contained optical unit which is also self-correcting (Fig. 23). It consists of only four groups of electrodes; two sets of eight poles and two sets of four poles. The first group (eight electrodes) provides both Φ_2 (which produces a large

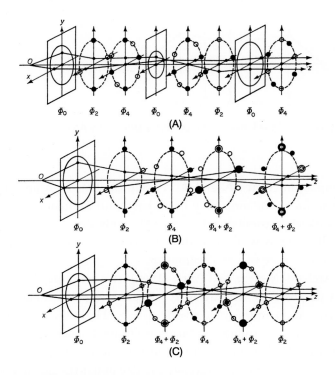

Figure 22 Three different systems suggested by Archard.

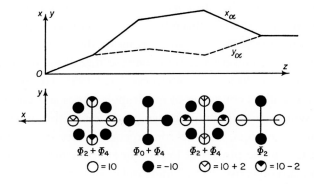

Figure 23 "Unconventional lens" of Archard, without a round lens.

astigmatism) and the corrector Φ_4 (corresponding to Seeliger's $O_{45°}$). The second group—four electrodes all at the same potential—is equivalent to a round lens, Φ_0, onto which is superimposed an octopole component (Φ_4).

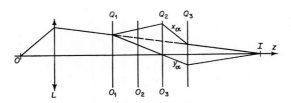

Figure 24 Arrangement of the lenses in the corrector studied by Archard, Mulvey, and Petrie.

The third group again creates Φ_2 and Φ_4. Finally, the fourth group, a pure quadrupole, transforms the astigmatic beam back into a rotationally symmetrical beam traveling parallel to the axis. The only disadvantage of the system is that the lens is weak: it is difficult to confine the four optical elements within the space of a few millimeters.

Recently, Archard, Mulvey, and Petrie (1960) have announced that one of the systems proposed by Archard is to be studied experimentally, with a view to correcting a round objective lens. Fig. 24 shows the general arrangement, which here produces an image at a finite distance. Q_1, Q_2 and Q_3 are the quadrupole elements producing $\Phi_2(z)$, and O_1, O_2, and O_3 are the octopole correctors which produce $\Phi_4(z)$. The potentials are chosen in such a way that the image I of the object O is formed in the same plane at the same magnification, whether the corrector system is excited or earthed. Only the change from a "direct" image to a "mirror" image is perceived, since the rays in the yOz-plane cross the axis inside the system when Q_1 and Q_3 are excited. The convergence is the same as that of the round lens.

(d) *The Glaser corrector.* In a series of reports dated 1955, Glaser studied the properties of a doublet consisting of quadrupole lenses and equivalent to a rotationally symmetrical system; the asymptotic foci of this doublet are automatically immersion foci, and the convergence is very strong. Glaser proposed to use this as the projective lens of an electron microscope, together with a standard objective; by including only two octopole lenses, the final image could be perfectly corrected, provided the distributions $\Phi_2(z)$ and $\Phi_4(z)$ were produced carefully enough, and in the proper positions along with axis.

Fig. 25 shows the ray paths through the microscope; $\Phi_2(z)$ is no longer merely an auxiliary element which enables the correction to be effected: it now takes part in the formation of the final image.

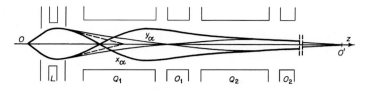

Figure 25 Electron trajectories in the *xOz* and *yOz* planes of Glaser's corrected system (schematic).

Although experiment (Septier, 1958) has shown that a quadrupole projective can provide an image of high quality, no corrected system exploiting this idea of Glaser's has, to my knowledge, been constructed.

(e) *Conclusion.* If, at this point, we examine all these proposals in detail, it is clear that in each, the axially symmetrical beam which emerges from the round objective is rendered astigmatic by the electrodes or lenses with quadrupole symmetry, and subsequently recovers its original symmetry with the aid of similar lenses. We are naturally led to the idea of suppressing the round lens altogether, as in one of Archard's systems, and of building optical combinations composed wholly of quadrupoles. This prompts us to ask two questions: Is it in fact possible to construct stigmatic systems without distortion (orthomorphic) using only quadrupoles, in which the foci are external and the focal length short (this would make them suitable for use as objective lenses, or as the final lenses of microanalyzers)? What would be the aperture aberration of such lenses, and could it be corrected?

We shall now try to reply to these two questions, in the light of recent work on quadrupole lenses.

5.3 Investigations of Quadrupole Lenses: Combinations Equivalent to Strongly Convergent Round Lenses

Examination of this problem reveals two possibilities. We can seek a system which will be stigmatic for an infinite number of pairs of conjugate points on the axis; in this case, the foci associated with the symmetry planes xOz and yOz must coincide, and the focal lengths F_x and F_y must be equal (the magnifications M_x and M_y are then equal, and there is no distortion). Alternatively, we can require stigmatic operation free of distortion for two particular points on the axis only.

(1). *Systems which are Stigmatic for Every Pair of Conjugate Points on the Axis.* In the simplest situation, the system must possess either a central

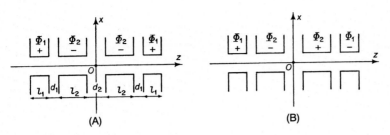

Figure 26 Different possibilities of obtaining optical systems equivalent to round lenses. (A) Mechanical symmetry and electrical symmetry with respect to the plane $z = 0$, (B) mechanical symmetry and electrical antisymmetry (the second part of the system is identical to the first one, but rotated through $\pi/2$ about the axis Oz).

plane of mechanical and electrical symmetry (of the excitations), normal to Oz, (Fig. 26(A)) or a central plane of mechanical symmetry and electrical anti–symmetry (see Fig. 26(B)). These systems must then satisfy two conditions: the x and y focal lengths must be equal, and the foci F_x and F_y must coincide.

The simplest arrangement of all is the symmetrical doublet, already mentioned above, which consists of two mechanically identical lenses with the same excitations, arranged so that the convergent plane of the second coincides with the divergent plane of the first. Such a doublet has been extensively studied by Glaser (1956), Reisman (1957), Septier (1958), and Dhuicq (1961).

Let us suppose each lens to be replaced by an ideal lens, for which the function $\Phi_2(z)$ or $B_2(z)$ is in the form of a rectangle of length L (the "equivalent length"); the excitation of the lens is characterized by β^2, defined by

$$\beta^2 = \frac{\mu_0 nI}{a^2} \left(\frac{2e}{M\Phi_0} \right)^{\frac{1}{2}}$$

in the magnetic case, and

$$\beta^2 = \frac{2\Phi_1}{a^2 \Phi_0}$$

in the electrostatic case (a denotes the radius of the inscribed circle which is tangent to the electrodes of the lens, $\pm\Phi_1$ the potentials applied to the electrodes, nI the number of ampère-turns per pole, Φ_0 the accelerating potential of the incident particles and M the mass of these particles). The

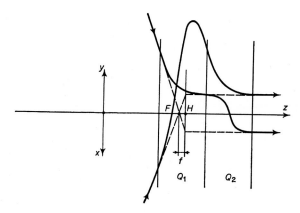

Figure 27 Trajectories and cardinal elements in a doublet equivalent to a system of revolution.

distance d, between the two lenses must satisfy the relation

$$\frac{L}{d} = -\frac{\beta L}{2}(\cot \beta L + \coth \beta L). \tag{38}$$

The focal length F is given by

$$\beta F = \frac{\sinh \beta L \cdot \cos \beta L + \cosh \beta L \cdot \sin \beta L}{\sinh^2 \beta L - \sin^2 \beta L}. \tag{39}$$

In the special case $d = 0$ (when the lenses are in contact), we must have $\beta L = \pi$; the system is then highly convergent. The trajectories which correspond to an incident beam parallel to the axis are shown in Fig. 27; the asymptotic foci lie within the lens, at about $L/3$ from the extremities, and the common focal length, $F_x = F_y = -\dfrac{a}{\pi}\dfrac{1}{\sinh^2 \pi}$, is of the order of $3 \times 10^{-2} L$, a few tenths of a millimeter.

Dhuicq and Septier (1959) have systematically tried to obtain strongly convergent optical systems with the foci external by combining three or four lenses. For a triplet L which always possesses both mechanical and electrical double symmetry about the central plane, the foci are always inside the system. With a quadruplet, still possessing mechanical symmetry, two situations are possible: electrical symmetry (Fig. 26(A)), and electrical antisymmetry (Fig. 26(B)).

The equations which must be solved in order to express the necessary excitations, $\beta_1 L_1$ and $\beta_2 L_2$, as functions of L_1, L_2 and the distances d_1, d_2

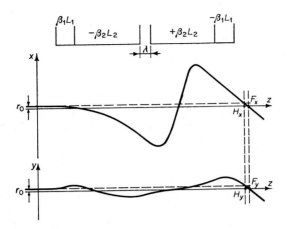

Figure 28 Trajectories in a strongly excited quadruplet equivalent to a short-focus round lens (Dymnikov and Yavor).

between the lenses, are so complicated that only with the aid of a computer can they be solved, even for special cases (as functions of the dimensionless ratios d_1/L_1 and d_2/L_2, for example) and using the approximate rectangular model described earlier.

For the case with electrical symmetry, Dhuicq has found that here again, the foci lie inside the system. He also set up the equations for the antisymmetrical situation, but did not explore them numerically. Dymnikov and Yavor (1964a, 1964b) have further examined these equations, however, and shown that here there are ways of constructing the desired system, for certain simple geometrical configurations. When, in particular, the lenses of each doublet are effectively in contact ($d_1 = 0$), the foci will be external, at a distance of about $0.4L_1$ from the entry and exit planes, for $L_2 = 4L_1$ with $d_2 \simeq L_1$; the excitations must satisfy $\beta_1 L_1 = 1.8$ and $\beta_2 L_2 = 4.3$, and it is found that $f' \simeq 0.01L$. Here again, the lens excitations are high, and the trajectories cross the axis several times within the lenses (Fig. 28). Furthermore, the trajectories in the plane which is initially convergent depart far from the axis, and their slopes, x', will be very steep; we may expect the aberration coefficients in the Ox, Oy, and 45° directions to be very different—in Eq. (33), we shall have $a \gg b \gg d$.

In these systems, the stigmatic orthomorphic operation is obtained by adjusting two potentials (or two currents).

(2). *Stigmatic Systems Free of Distortion for only Two Points on the Axis.* At least three lenses are necessary to obtain these working conditions. One

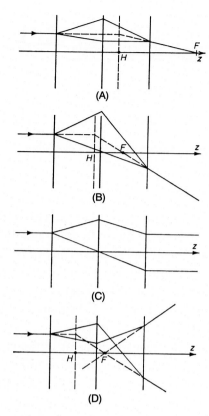

Figure 29 Several examples of triplets equivalent to round lenses (thin lens approximation).

example has already been found, the corrector of Archard, Mulvey, and Petrie, but there the lens was weak.

An examination by Deltrap (1964a) has shown that a great many such triplets can be obtained (even when the lenses are equidistant), some of which may be quite strongly convergent ($F \simeq L$). A few examples are shown in Fig. 29, which all correspond to an incident beam parallel to the axis. Apart from the lens separations, three electrical parameters are at our disposal, with which the desired conditions can be fulfilled.

With four lenses, the number of combinations is still greater, and it is then possible to obtain strongly convergent systems.

We have studied in detail only one system of this type: an electrostatic triplet with unit magnification (Fig. 30). The triplet consists of two short

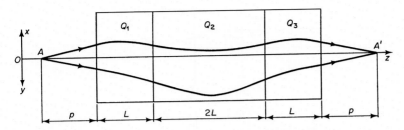

Figure 30 Symmetrical electrostatic triplet with unit magnification.

lenses of length L, Q_1 and Q_3, which have the same excitation, and a central lens Q_2 of length $2L$, having a different excitation. For simplicity, we chose the distance between the lenses to be zero, and the radius of the inscribed circle was the same for all three lenses. The first order calculations are relatively straightforward, and only two potentials have to be varied to fulfill the conditions of stigmatic and orthomorphic imagery between the selected pair of conjugate points A and A' (which are symmetrical about the center, O', of the triplet).

The principal planes and the foci are different in the two planes xOz and yOz, but the complete system can be considered to be equivalent to a thin lens placed at O', with focal length $F_x = F_y = \frac{1}{2}AO' = \frac{1}{2}A'O' = F$.

Such a triplet might well be used to form an electron probe from a very small crossover, itself produced by a quadrupole doublet or even more simply, by a point cathode (Crewe, 1965).

5.4 The Aperture Aberrations of Quadrupole Lenses

(1). *General Expressions: Calculations.* If the equations of motion are calculated taking the third–order terms into account, then for a purely quadrupolar field (by which we mean a field in which there are no terms of higher than first-order apart from those describing the fringe fields at the extremities), we find

$$B_X = K(z)Y - \frac{1}{12}K''(z)Y(3X^2 + Y^2)$$
$$B_Y = K(z)X - \frac{1}{2}K''(z)X(3Y^2 + X^2) \qquad (40a)$$
$$B_z = K'(z)XY$$

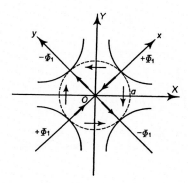

Figure 31 Schematic transverse section of a quadrupole lens.

in the magnetic case, and

$$E_x = K(z)x - \frac{1}{6}K''(z)x^3$$

$$E_y = -K(z)y + \frac{1}{6}K''(z)y^3 \tag{40b}$$

$$E_z = \frac{1}{2}K'(z)\left(x^2 - y^2\right)$$

in the electrostatic case.

The axes OX, OY, Ox, and Oy are defined in Fig. 31; $K(z)$ represents the gradient function along Oz: (see Fig. 32)

$$K(z) = \frac{2\Phi_1}{a^2}k(z) \quad \text{(electrostatic)}$$

$$K(z) = \frac{2\mu_0 nI}{a^2} - k(z) \quad \text{(magnetic)} \tag{41}$$

If the lenses are very short (the mechanical length $l \leqslant$ diameter $2a$ of the central opening), the gradient at the center, K_0, is given by

$$K_0 = s\frac{2\Phi_1}{a^2}$$

or

$$K_0 = s\frac{2\mu_0 nI}{a^2} \tag{42}$$

in which s is a correction factor smaller than unity; it is a function of the ratio l/a.

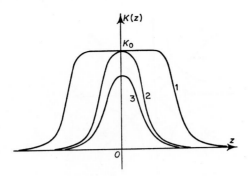

Figure 32 Appearance of the characteristic gradient function $K(z)$ along Oz. Curve $1 =$ long lens $(l > 2a)$; curve $2 =$ relatively short lens $(l \simeq 2a)$; curve $3 =$ very short lens $(l < 2a)$. l is the mechanical length of the poles (or electrodes) and $2a$ the diameter of the useful aperture (see Fig. 31). For a very short lens, the maximum gradient at $z = 0$ is less than the theoretical value K_0.

The terms in K' and K'' are only present within the fringe fields; at the center of the lens, they vanish provided the latter is long enough.

To take into account the slope of the trajectories within the lens, we must write

$$\frac{dz}{dt} \simeq v\left[1 - \left(x'^2 + y'^2\right)\right]^{\frac{1}{2}} \tag{43}$$

(v is the total velocity of the particle at the point in question).

Finally, the equations of the trajectories are obtained to the third-order approximation in the planes xOz and xOy: $x^{(3)}(z)$ and $y^{(3)}(z)$.

The transverse aberrations, ε and η, can be calculated rapidly at every point along the trajectories by regarding the aberrations as small perturbations of the first order trajectories, $x^{(1)}(z)$ and $y^{(1)}(z)$. Thus

$$\varepsilon = x^{(3)}(z) - x^{(1)}(z)$$
$$\eta = y^{(3)}(z) - y^{(1)}(z). \tag{44}$$

In the plane OX, in the magnetic case for example, ε is found to be the solution of the equation:

$$\varepsilon'' + k(z)\varepsilon = -\beta^2\left[\underbrace{k(z)X\left(\frac{3X'^2 + Y'^2}{2}\right) - k(z)X'Y'Y}_{A} - \underbrace{k'(z)XYY'}_{B}\right.$$
$$\left.\underbrace{-\frac{k''(z)}{12}X(X^2 + 3Y^2)}_{C}\right]. \tag{45a}$$

In the plane Ox, in the electrostatic case, we find

$$\varepsilon'' + k(z)\varepsilon = -\beta^2 \left[\underbrace{k(z)\left(x'^2 + y'^2\right)}_{A} + \underbrace{\beta^2 k^2(z)x\left(x^2 - y^2\right)}_{A'} - \underbrace{\frac{1}{2}k'(z)x'\left(x^2 - y^2\right)}_{B} \right.$$
$$\left. - \underbrace{\frac{k''(z)}{6}x^3}_{C} \right]. \tag{45b}$$

Aberration terms of various origins appear in these equations: *The terms A* are due to the slope of the trajectories, and are always present, even if $k(z) = $ constant. The stronger the excitations of the lenses, the more important will they be.

The term A' appears only in electrostatic lenses, and is due to local variations in the energy, which arise from the changes in the potential $\Phi(x, y, z)$ within the lens.

The terms B describe the effect of the longitudinal field B_z (or E_z), and only appear near the ends of the lenses.

Finally, *the terms C* correspond to the round lens term, since they are functions of $k''(z)$; they only exist within the fringe fields.

Furthermore, if the lens is not ideal—if, that is, the gradient in the central zone is not rigorously constant along a radius—supplementary terms appear. In general the gradient decreases as we move away from the axis of a real lens L in the planes XOz and YOz, which are midway between the pole-pieces. This decrease is physically equivalent to a reduction of the "equivalent length", $L(r)$, in the outer regions, and hence to a reduction of the local convergence of the lens; this creates aberration terms different in sign from those of third order. It must not be thought that the third-order aberrations of the lenses could really be corrected by means of these supplementary aberrations, however, for they are not of the same order in r; we may nevertheless anticipate some local compensation.

For an incident beam parallel to the axis, the transverse aberration in the Gaussian image plane can be written in the form

$$\Delta x = C_1 \alpha^3 + C_2 \alpha \beta^2$$
$$\Delta y = C_2 \alpha^2 \beta + C_3 \beta^3. \tag{46}$$

α and β are the semi-aperture angles in image space, in the planes xOz and yOz. The longitudinal aberration is then given by

$$\Delta z_x = C_1\alpha^2 + C_2\beta^2$$
$$\Delta z_y = C_2\alpha^2 + C_3\beta^2. \tag{47}$$

In weakly excited lenses, the terms arising from the slope of the trajectories are negligible; we have verified this theoretically by integrating the preceding equations first with, and then without the terms A.

The terms C, which are due to the fringe fields, can be reduced by using long lenses, or by producing particularly suitable distributions $k(z)$; for example, $k''(z)$ could be canceled by modifying the shape of the poles in the Oz-direction.

In certain situations, it might be valuable to correct the transverse aberration; if the image of a point is to be formed at the entry slit of a spectrograph, only C_1 for example, need be corrected if the first focal line produced by the lens is employed. We have measured these transverse aberrations for weak electrostatic and magnetic lenses ($F \geqslant L$). The experimental results are in good agreement with the corresponding calculations, and we may conclude that the equations given above are accurate enough for the determination of the third order aberrations of a quadrupole system; for $k(z)$ we use some simple analytic function which gives as close a description as possible of the real curve obtained by experiment. For short lenses, a bell-shaped curve of the type

$$k(z) = \left(1 + \left(\frac{z}{b}\right)^2\right)^{-2} \tag{48}$$

is excellent. If the lens is long, the function $k(z)$ has a central plateau (cf. Fig. 32), so that we must consider two half-bells separated by a region in which $k(z) = 1$. To calculate the aberrations of systems containing several lenses, the equations have to be integrated from one end of the system to the other.

Alternatively, we can use a more rapid approximate method (Deltrap, 1964a) in which the aberrations of each of the lenses, transmitted through the remainder of the system, are superimposed in the final image plane. Only the aberrations of the different lenses used need then be known.

Finally Hawkes (1962, 1963, 1965a), extending Sturrock's (1951, 1955) perturbation characteristic function analysis to non-rotationally symmetrical systems, gives a unified representation of the electron optical properties

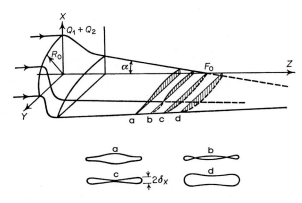

Figure 33 The astigmatic doublet: cross-section of a hollow cylindrical incident beam, near the first focal line. If the figures (a) (b) (c) (d) appear on a fluorescent screen in the order (a) to (d) as the convergence increases, the aberration is called "negative"; (magnetic doublet). It is called "positive" if the figures appear in the order (d) to (a) (symmetrical electrostatic doublet).

of electrostatic, magnetic, or mixed orthogonal systems, in which round and quadrupole lenses are only particular cases.

Using the recent results obtained by Hawkes (1965b) it will now be possible to calculate rigorously the aperture aberration coefficients by numerical integration of definite integral expressions, containing only the potential (or magnetic field) function, and the expressions for two particular first-order trajectories.

(2). *The Magnitudes of the Aperture Aberrations: Astigmatic Systems.* We have measured (Grivet & Septier, 1960; Septier, 1961) the transverse aberration of quadrupole doublets, consisting of two lenses for which $L = 20$ cm, $a = 4$ cm, separated by a distance $D = 30$ cm, under astigmatic conditions; this aberration is characterized by the maximum half-width, δ_x, of the line focus which is deformed by third order aberration (Fig. 33), and by the factor $\tau_x = \delta_x / R_0$ (where R_0 is the radius of a hollow cylindrical incident beam which enters the doublet parallel to the axis). The quantity τ_x, is proportional to the constant C_1 corresponding to the Gaussian image plane. Even though the real values of the coefficients are not obtained with this method, we have been able to compare the optical properties of various magnetic and electrostatic lenses.

The main results are set out in Figs. 34 and 35; the former corresponds to a magnetic doublet. For a given value of R_0, δ_x increases very

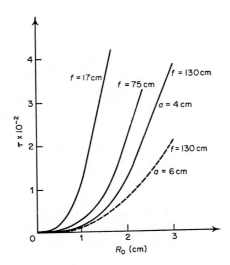

Figure 34 Astigmatic magnetic doublet: transverse aberration factor $\tau_x = \delta_x/R_0$ ($L = 20$ cm, $a = 4$ cm, $D = 30$ cm) as a function of R_0, for incident hollow beam parallel to the axis, and different values of the focal length. The dotted curve corresponds to $f = 130$ cm and $a = 6$ cm.

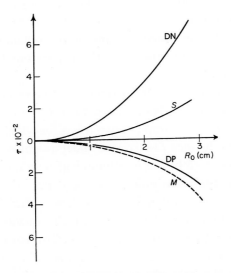

Figure 35 Astigmatic electrostatic doublet: aberration factor $\tau_x = \delta_x/R_0$ as a function of R_0 for $f = 130$ cm. Curve S = symmetrical excitation of the lenses $(+\phi_1, -\phi_1)$. Curve D.P. = asymmetrical positive excitation $(+2\phi_1, 0)$. Curve D.N. = asymmetrical negative excitation $(0, -2\phi_1)$. Dotted curve M = magnetic doublet with the same convergence.

rapidly with the convergence. On increasing the internal radius from four to six centimeters, however, an appreciable improvement is found, which is doubtless due to the smaller effect of the terms in r^5. If, on the other hand, $C^{(3)}$ and $C^{(5)}$ are calculated, where

$$\tau = C^{(3)} \left(\frac{R_0}{f} \right)^2 + C^{(5)} \left(\frac{R_0}{f} \right)^4$$

the following results are obtained:

F/L	0.8	3.75	6.5
$C^{(3)}$	3	14	25
$C^{(5)}$	1.7×10^2	2.3×10^4	10^5

The coefficients $C^{(3)}$ and $C^{(5)}$ decrease rapidly as the convergence is increased.

We now consider an electrostatic doublet (Septier & Van Acker, 1961), geometrically identical to the magnetic doublet (Fig. 35); the curves "S" show the aberration of a symmetrically excited doublet (in which adjacent electrodes are held at potentials $+\Phi_1$). Comparing these with the magnetic case, which corresponds to the broken curve "M", we see that the aberration is virtually the same, but of opposite sign. If the doublet is excited asymmetrically, two electrodes of each lens being earthed and the other two held at a positive potential $+2\Phi_1$, the aberration is still approximately the same, but has the same sign as in the magnetic case. If finally, the excitation is asymmetrical and negative (0 and $-2\Phi_1$), the aberration is considerably increased. The sign of the aberration can therefore be reversed by introducing a degree of asymmetry into the electrode excitations (see Fig. 36).

If the shape of the poles is altered, without destroying the quadrupole symmetry, the aberration varies between wide limits (Fig. 37), and may even change sign. We recognize the influence of the variation of the equivalent length, $L(r)$, of the lens with r, already mentioned (see Section 5.4), especially in curve No. 6.

More recently, Deltrap (1964a) has used the grid shadow method proposed by Deltrap and Cosslett (1962) to measure the coefficients C_1 and C_2 of individual short magnetic lenses in which the $B(z)$ curves have no central plateau; in these lenses, the pure quadrupole distribution in two dimensions does not exist.

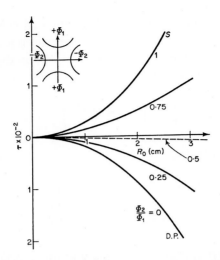

Figure 36 Variation of the aberration factor τ_x with the degree of asymmetry in the excitation $(+\phi_1, -\phi_2)$; the transverse aberration is practically corrected when $\phi_2/\phi_1 \simeq$ 0.5.

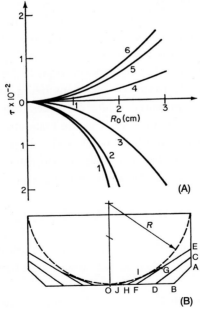

Figure 37 Influence of the shape of the cross-section on the aberration factor τ_x. Curves 1...6 correspond respectively to the profiles OBA, ODC, OFE, OHGE, OJIGE, and a circle of radius $R = 1.15a$. The profile OJIGE represents to a very good approximation the theoretical hyperbola.

As the convergence increases, the coefficients C_1/L and C_2/L decrease very rapidly; when L is increased for a constant value of a, they decrease very slowly. At high values of the convergence ($F/L \leqslant 2$), they are comparable with those of round lenses; when $L = F$, for example, one finds $C_1 \simeq F/3$ and $C_2 \simeq F/3$.

If several of these lenses are combined to form a triplet or quadruplet, the aberration coefficients increase very rapidly, and for systems with external foci they are very large indeed. (In the systems studied by Deltrap, the corresponding focal lengths are relatively long.) For the triplet shown schematically in Fig. 30, for example, we have $C_1 \simeq F/2$, $C_2 \simeq 9F$, and $C_3 \simeq 60F$ with $F/L \simeq 2.5$.

For systems which arc equivalent to round lenses and have immersion foci, however, the coefficients arc considerably smaller, and become comparable with those of round lenses. For a magnetic doublet, Reisman (1957) finds

$$C_1 = 5F, \qquad C_2 = 2F, \qquad C_3 = 25F, \quad \text{with } F/L \simeq 0.05 \ (F = 2 \text{ mm});$$

for a triplet (Fig. 29(B)), Deltrap finds

$$C_1 = -0.25F, \quad C_2 \simeq 0.2F, \quad C_3 \simeq 4F, \quad \text{with } F/L \simeq 2/3 \ (F = 20 \text{ mm}).$$

About systems of four lenses equivalent to round lenses, with short focal lengths and external foci, no data are at present available. We might expect the coefficients to be similar to those of the preceding systems; a programme of calculations which is at present in progress will elucidate this point.

We have also measured the coefficients of the symmetrical electrostatic triplet with long lenses, described above (Septier, 1962, 1963; Septier & Dhuicq, 1964). The results are as follows:

$$C_1 \simeq F, \quad C_2 \simeq 15F, \quad C_3 \simeq 140F, \quad \text{with } F/L_2 \simeq 1.$$

(3). *Recent Attempts to Correct Quadrupole Systems.* Systems of quadrupole lenses can in theory be corrected easily, by incorporating local perturbations with octopole symmetry, since the quadrupoles do not possess rotational symmetry.

(i) The first-order optical properties of an astigmatic electrostatic doublet (Septier, 1962; Septier & Van Acker, 1961) are practically constant if the excitations of the electrodes are made asymmetrical ($+\Phi_1$ on one pair,

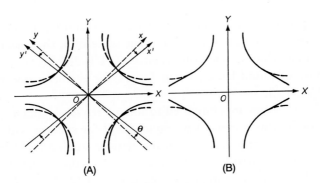

Figure 38 Two practical ways of introducing octopole correction in a quadrupole lens. (A) Rotation of the poles through an angle θ about Oz. (B) Deformation of the poles near the plane ZOX.

$-\Phi_2$ on the other), provided that the total potential difference, $(\Phi_1 + \Phi_2)$ remains unaltered during the operation.

The potential $\Phi(r, \theta, z)$ can easily be shown to consist of three terms: the dominant quadrupole term, the octopole term Φ_4, and a round lens term, Φ_0; thus, $\Phi = \Phi_0 + \Phi_2 + \Phi_4$.

(ii) Another way of introducing the octopole correction is to relinquish the quadrupole symmetry of the poles, in one of the ways shown in Fig. 38: we can either rotate two of the poles through an angle θ (Markovich & Tsukkerman, 1961) or modify the shapes of the poles locally, always retaining the two symmetry planes (Reisman, 1957).

(iii) The most flexible and the most general method is obviously to include octopole lenses in the system; the correcting action of such a lens on the aberration figure of an astigmatic system is illustrated in Fig. 39; a considerable reduction of the overall aberration of a stigmatic doublet with unequal magnifications M_x and M_y can be obtained by means of a single octopole (Fig. 40(C)), since C_1 and C_3 are so very different in this case ($C_3 \simeq 100 C_1$).

To render the corrected system less bulky, the quadrupoles and octopoles might be combined; this could be done either by introducing additional electrodes within the quadrupoles (Septier & Dhuicq, 1964; Fig. 41), or by employing mixed magnetic lenses with eight poles (Deltrap, 1964a, 1964b) in which the shapes of neighboring poles are designed in such a way that they reproduce the shapes of the poles of a quadrupole lens to a good approximation (Fig. 42). The third order aberration co-

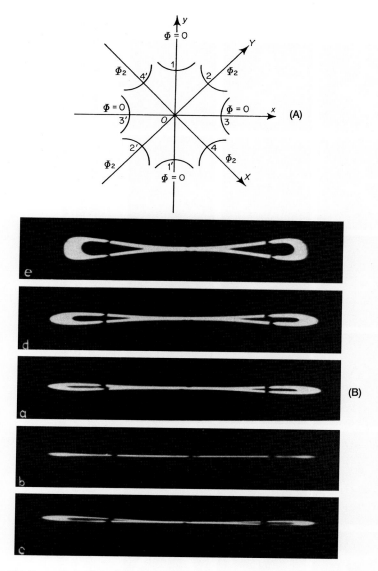

Figure 39 Correction of the transverse aberration of an astigmatic doublet using an octopole lens. (A) Schematic diagram of an octopole lens; (B) Photographs: (a) initial aberration figure, (b) and (e) negative potentials $(-\Phi_2)$ applied to the electrodes 2–2′ and 4–4′ of the octopole lens; the aberration decreases (case (b)) and changes sign (case (c)); (d) and (e) positive potentials $(+\Phi_2)$ on the electrodes 2–2′ and 4–4′. The aberration increases. The electrodes 1–1′ and 3–3′ are maintained at $\Phi = 0$.

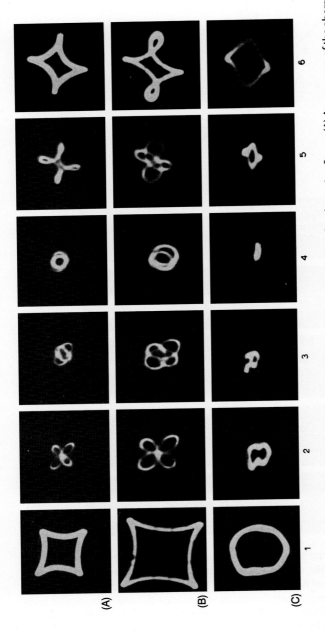

Figure 40 Pseudo-stigmatic doublet ($M_x \neq M_y$): influence of the octopole potential on the aberration figure. (A) Appearance of the aberration figure on the fluorescent screen, the excitation increasing from 1 to 6—octopole excited in the wrong sense ($+\Phi_2$ on the electrodes 2–2′ and 4–4′); (B) the same figures, with the octopole not excited; (C) the same figures, with the correct potential ($-\Phi_2$ applied to the octopole electrodes 2–2′ and 4–4′).

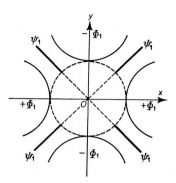

Figure 41 Modified quadrupole lens (Septier and Dhuicq). Additional electrodes, supplied with a potential $\psi_1 \neq 0$ and situated in the planes *ZOX* and *ZOY* introduce octopolar correction.

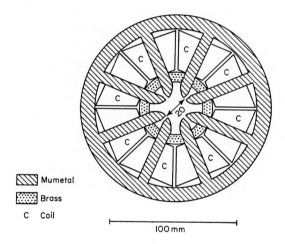

Mumetal
Brass
C Coil

100 mm

Figure 42 Mixed magnetic lenses with eight poles (Deltrap). This lens can give quadrupolar and octopolar fields simultaneously, by proper adjustment of the values of the currents in the coils C.

efficients of these mixed lenses are very close to those of ordinary lenses ($C_1 \simeq C_2 \simeq 0.25F$).

Attempts have also been made to correct systems of three and four lenses. Experiment has shown that the quality of the spot produced by an electrostatic triplet of unit magnification can be considerably improved by applying the excitations to the three lenses asymmetrically (the first and third lenses are always excited identically, which means that there are only two octopole effects to perform the correction). The aberration cannot,

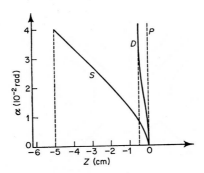

Figure 43 Longitudinal aberration Δz in a symmetrical electrostatic triplet as a function of the semi-angular aperture α (see Fig. 30); $p/L \simeq 0.75$.

however, be canceled in this way, since for large asymmetries, the round lens component Φ_0 introduces a further third order aberration which is not at all negligible. Fig. 43 shows the overall longitudinal aberration, Δz, measured on an optical bench using a very large triplet; the position of the crossover of the beam is located on a fluorescent screen as the angular aperture α is varied. The maximum overall improvement obtained is about 10 for $\alpha \simeq 4 \times 10^{-2}$ radians.

Using the mixed lenses shown in Fig. 41, it has proved possible to change the sign of the variations of Δz, for particular values of the potentials ψ_1 and ψ_2 applied to the corrector electrodes of Q_1, Q_3, and Q_2 respectively. This measuring technique enables us to define a mean coefficient, $\overline{C}_3$:

$$\Delta z = \overline{C}_3 \alpha^2$$

with $\overline{C}_3 \simeq 30F$ for an uncorrected symmetrical triplet.

It is clear that although it may be possible to obtain large aberrations of opposite sign (Fig. 44), the aberration cannot be completely canceled for a beam with angular aperture α of the order of 10^{-2}; this is due to the presence of higher order terms, which explains the curious appearance of the curves obtained close to the optimum correction conditions (Fig. 45). At best, $\overline{C}_3 \simeq F$. As we shall see later, however, these terms are undoubtedly due to the mechanical shortcomings of the system, and could be considerably reduced by assembling the latter more carefully. Here again, the correction has been effected by only two octopoles, since the system is symmetrical about a central plane; complete correction could be achieved only by inserting a third octopole between the lenses.

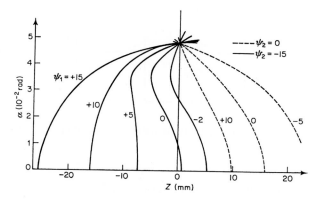

Figure 44 Attempts to correct the aberration (symmetrical electrostatic triplet $p/L \simeq$ 0.75). Potential ψ_1 is applied to the additional electrodes of quadrupoles Q_1 and Q_3; potential ψ_2 corresponds to the electrodes of Q_2. The values of ψ_1 and ψ_2 are given in arbitrary units.

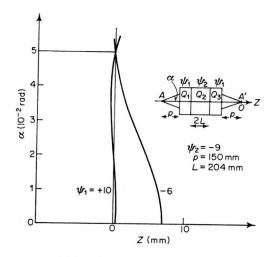

Figure 45 Correction of the aberration: longitudinal aberration Δz as a function of the semi-aperture α, better adjustment than in Fig. 44.

Deltrap (1964a) has attempted to correct a system of magnification less than unity, which is very similar to the type of system used in micro-probe devices: it consists of four mixed lenses followed by a round lens. The reason for this choice is simple: when two lenses, L_1 (of magnification M_1 near to unity) and L_2 (of magnification M_2, much less than unity), are used, it is

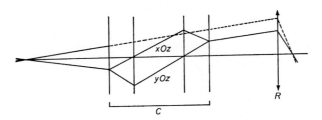

Figure 46 Trajectories in Deltrap's corrector (schematic): R = round objective lens; C = corrector (four mixed magnetic lenses).

easy to show that the overall aberration coefficient, C_{tot}, is given by

$$C_{tot} \simeq C_1(M_2)^4 + C_2.$$

Even if C_1 is very large (and we have seen that this is the case for very many weakly convergent systems of quadrupoles), the spherical aberration of the combination is only slightly greater than that of the final lens, L_2.

It will therefore be advantageous to choose a lens with very low aberration as L_2: either a round lens, or a combination of quadrupoles equivalent to a round lens and strongly convergent would be suitable (provided that the latter kind of combination with external foci can be shown to have relatively small aberration coefficients). The same would be true for an electron microscope objective followed by a corrector with magnification equal to, or close to unity.

The system suggested by Deltrap is shown schematically in Fig. 46 and the design is intended to facilitate the final adjustment of the octopoles. Two of the mixed lenses lie in planes in which the beam forms a real focus, as in Scherzer's design. In the system which was in fact constructed, the final magnification is only 1/8 (instead of 1/50 to 1/100, which is used in real systems). Fig. 47 shows curves obtained by the grid shadow method which describe the longitudinal aberration Δz in the two planes xOz, yOz and the plane at 45° to these. The total residual aberration, of fifth order, is such that

$$C_5 \simeq 20\overline{C}_3,$$

in which $\overline{C}_3$ denotes the aberration of the round lens alone ($\overline{C}_3 \simeq 80$ mm).

These results are extremely encouraging, therefore, but their real validity still remains to be demonstrated: a *real* test with the real dimensions needs to be performed, of the kind performed by Seeliger and Möllenstedt. Thanks

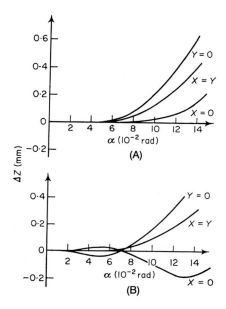

Figure 47 Residual longitudinal aberration Δz in Deltrap's corrector, as a function of the angle, for different states of the correction. The reduction of Δz is not exactly the same for the different terms (planes $X = 0$, $Y = 0$, and $X = Y$).

to the analyses of Archard (1954a), Meyer (1961), and Deltrap (1964a), however, it is now possible to make a good estimate of the limitations of this kind of system. We shall examine this question in the next section.

6. THE ULTIMATE PERFORMANCE OF CORRECTED SYSTEMS

6.1 Additional Aberrations Arising from Mechanical Defects

A naive line of reasoning might lead one to hope that if the third-order aberration were corrected, the performance of the system (the resolving power of a microscope objective, or the current in a microprobe of given diameter) could be determined by considering the residual fifth-order aberration of the lens being corrected (or of the complete quadrupole system). In fact, the electrodes present in the corrected combinations are necessarily so numerous that the mechanical errors in the construction of both quadrupoles and octopoles will play an important role; it is therefore necessary to determine, in an approximate fashion at least, the orders of magnitude of the tolerances which will be required if we are to obtain a substantial improvement in performance, in comparison with that of the lens being corrected.

In the case of a round objective, the theoretical resolving power, in which both the third order aberration and the diffraction are taken into account, is given by

$$d_{th} = K\lambda(C_3/\lambda)^{\frac{1}{4}} \quad \text{with } K \simeq 0.5;$$

λ denotes the wavelength associated with particles accelerated through a potential Φ.

For the electrostatic objective of medium quality used by Seeliger ($C_3 = 6$ cm), and 40 kV electrons, we find

$$d_{th} \simeq 10 \text{ Å}.$$

When $C_3 = 0$, we have

$$d_{th} = K_5\lambda(C_5/\lambda)^{1/6}$$

so that $d_{th} = 2$ Å for $C_5 = 15$ m, a fivefold improvement.

The residual ellipticity of the openings of this round objective must be less than two microns, for this improvement to be attainable; this is still mechanically feasible.

For a magnetic objective, the improvement will be of the same order of magnitude if C_3 can be canceled.

Let us now consider the effects of small mechanical errors on a system of round lenses and octopoles. The ideal potential distribution, described by

$$\Phi(r, \theta, z) = \Phi_0(r, \theta, z) + \Phi_2(r, \theta, z) + \Phi_4(r, \theta, z) \tag{49}$$

will contain additional perturbation terms. Neglecting terms of first degree in r, which correspond to a deflection and create no new aberrations, we obtain

$$\Phi(r, \theta, z) = \Phi_0 - \frac{1}{4}\Phi_0''r^2 + \frac{1}{64}\Phi_0''''r^4$$
$$+ \frac{1}{4}\cos 2\theta\left(\Delta r^2 - \frac{1}{24}\Delta''r^4\right)$$

$$+ \frac{1}{2} \sin 2\theta \left(\Phi_2 r^2 - \frac{1}{12} \Phi_2'' r^4 \right)$$

$$+ \frac{1}{3} \cos 3\theta \left(G_1 r^3 \right) - \frac{1}{3} \sin 3\theta \left(H_1 r^3 \right)$$

$$+ \cos 4\theta \left(\Delta_1 r^4 \right) + \sin 4\theta \left(\Phi_4 r^4 \right)$$

$$+ \text{further terms} \tag{50}$$

in which Δ, G_1, H_1, and Δ_1 are all functions of z.

New terms in r^2, r^3, and r^4 have appeared, which lead to aberrations of first-order (ordinary astigmatism), second-order (second-order astigmatism), and third-order respectively.

The first-order effects can be eliminated with the aid of a lens of the Stigmator type (Bertein, 1947, 1948; Rang, 1949); such lenses are regularly used in electron microscopy. The third order terms combine with the third-order aberrations to be corrected (we have seen earlier that they can be used deliberately to introduce an octopole corrector effect). If the second order effects are too troublesome, however, they must be corrected with the aid of an assembly of supplementary six-pole and two-pole lenses (Meyer, 1961).

If, furthermore, the poles of the quadrupoles are not hyperbolic in cross-section, the expansion above will contain terms in $r^6 \sin 6\theta$, and these will create an additional fifth order aberration.

Deltrap has calculated the possible improvement in the properties of an optical system for pre-determined values of the second- and third-order aberrations produced in this way. For a microscope objective the improvement parameter, t, will be defined thus:

$$t_2 = (C_3/\lambda)^{\frac{1}{4}} / (C_{(2)}/\lambda)^{\frac{1}{3}} \quad \text{for the second order}$$
$$t_3 = (C_3/C_{(3)})^{\frac{1}{4}} \quad \text{for the third order}$$
$$t_4 = (C_3/\lambda)^{\frac{1}{4}} / (C_5/\lambda)^{\frac{1}{6}} \quad \text{for the fifth order.}$$

$C_{(2)}$ and $C_{(3)}$ are measures of the parasitic second and third-order aberration respectively; C_3 is the third-order aberration constant of the lens R to be corrected (or of the complete quadrupole system which might perhaps replace it); and C_5 is the fifth-order aberration constant corresponding to R and the quadrupoles.

For a probe, it is better to discuss the current, I_{max}, which can be directed into a probe of given diameter, d. We have

$$I_{max} = K_3 C_3^{-2/3} \Phi j_0 T^{-1} d^{-8/3}$$

Table 1 Values of parasitic and fifth-order aberration coefficients

		t-objective$(\Phi = 100 \text{ kV}, C_3 = 2 \text{ mm})$	*t*-probe$(\Phi = 20 \text{ kV}, C_3 = 20 \text{ mm})$	
			$d = 0.1\ \mu$	$d = 1\ \mu$
$C_{(2)}/C_3$	10^{-2}	0.9	1.7	3.7
	10^{-3}	1.9	17	37
	10^{-3}	4	170	370
$C_{(3)}/C_3$	10^{-1}	1.8	4.6	4.6
	10^{-2}	3.2	22	22
	10^{-3}	5.6	100	100
C_5/C_3	1	5.3	26	14
	10	3.6	11	5.6
	100	2.5	4.1	2.3

where Φ is the accelerating potential, j_0 is the current density emitted by the cathode of the gun and T is the working temperature of the cathode. We then find

$$t_2 = C_3^{2/3} d^3 / C_{(2)} d^{8/3}$$
$$t_3 = (C_3/C_{(3)})^{2/3}$$
$$t_4 = C_3^{2/3} d^{12/5} / C_5 d^{8/3}. \tag{51}$$

In Table 1, the possible improvement parameters are listed, for a number of values of $C_{(2)}/C_3$, $C_{(3)}/C_3$, and C_5/C_3.

For an objective, we can see that the aberration which is likely to be the most troublesome is the second-order one. For a probe, on the other hand, the fifth-order aberration will prevent any substantial increase of the current.

With an isolated quadrupole, the accuracy in machining and assembly, corresponding to a ratio $\varepsilon(4\theta)/\varepsilon \leqslant 10^{-2}$ of the third-order aberration to the normal aberration to be corrected, will be about 2×10^{-2} mm with an internal radius $a_0 = 20$ mm and $F/L < 2$.

The relative importance of the second order aberration becomes greater if beams with very small angular apertures are employed; for $\alpha = 0.01$ radians, for example, $\theta(3\theta)/\varepsilon = 0.1$.

These orders of magnitude are unaffected when several quadrupoles are combined, and when the system contains both quadrupoles and octopoles.

Archard (1954a) also estimated the tolerances of the correctors which he proposed; an accuracy of the order of 10^{-2} mm in the positioning of the plate electrodes of his corrector lenses would be required.

The additional aberrations in 4θ decrease as the convergence is increased, whereas $\varepsilon(3\theta)/\varepsilon$ increases. A much higher degree of mechanical precision would therefore be necessary for a short focal length microscope objective working with a beam of very small angular aperture than for an electron probe lens, in which both F and α are larger; the various electrodes would have to be machined and assembled to within a few microns, just as in round objective lenses.

Translations and rotations of the electrodes of the octopoles likewise create aberrations of the same order of magnitude as those arising from the quadrupoles; as much care will have to be devoted to the corrector elements as to the quadrupole lenses, therefore.

An examination of all these conditions justifies Deltrap's final choice. By grouping the quadrupoles into a corrector system with low magnification, and combining this system with a strongly convergent round lens, R, the relative magnitudes of the aberrations arising from mechanical imperfections are themselves diminished, just as is the ordinary third-order (and fifth-order) aberration, as we have already mentioned. Nevertheless, the aberrations arising from the octopoles will not be reduced in this system, since these elements do not participate in the first-order image formation.

In a corrected system like that of Deltrap, which is intended for use with an electron probe device, we can conclude that tolerances of the order of a few hundredths of a millimeter would be necessary, with $a = 10$ mm, to obtain a ten or twentyfold improvement in the current in a microprobe of diameter 0.1 μ; for a microscope objective, the improvement would be in the neighborhood of two.

It is to be noticed that the tolerances vary in direct proportion to the scale; if the dimensions of the system were reduced by a factor five, the tolerances would fall to about a micron, which seems to be at the limit of practical possibilities. It would be more favorable to reduce the dimensions of the round lens without altering those of the quadrupole and octopole elements.

Finally, Meyer (1956) believes that the fifth-order aberration can be reduced, the third order aberration remaining canceled, by a judicious choice of the corrector elements. Deltrap's measurements appear to confirm that the final fifth-order aberration is in fact smaller than that of the round lens alone; he finds that after correction, $C_5 \simeq 20C_3$.

6.2 Other Limitations

When the third-order aberration has been eliminated, we must follow the example of Meyer (1961) and consider the limitations imposed by other types of aberration.

(1). *Chromatic Aberration.* This is due to fluctuations of the accelerating potential and of the currents in magnetic lenses. In view of recent progress in the electronics of stabilized voltage and current supplies, it now seems possible to attain a relative stability of $\Delta\Phi/\Phi \simeq \Delta i/i \simeq 10^{-6}$ during the period necessary for adjustment or observation. With electrostatic lenses, the situation is considerably more favorable, provided that all the potentials are taken from potentiometers supplied by a single generator.

It might also be possible to install a chromatic aberration corrector, of the kind described recently by Dupouy, Perrier, and Trinquier (1963, 1964). This is an electronic device which converts the fluctuations of the gun potential into variations in the lens currents; in this way, an important gain in resolution can be achieved.

(2). *The Space Charge of the Beam.* This space charge makes the beam diverge, especially near the crossover where the charge density is high. Instead of a point image, we shall find a spot of radius (Meyer, 1958)

$$\Delta r_0 = 4.8 \times 10^4 \frac{j_0 R_0^2 f}{\Phi_0^{3/2}} \sigma(\alpha) \tag{52}$$

in which $2R_0$ is the diameter of the beam, j_0 is the current density passing through the object, f is the focal length of the objective, Φ is the accelerating voltage and $\sigma(\alpha)$ is a function of the angular aperture.

For $j_0 = 1$ A cm^{-2}, $\Phi_0 = 40$ kV, we have $\Delta r_0 = 4$ Å. Fortunately, however, this aberration can be compensated by slightly over-correcting the third order aberration.

6.3 Correction in Practice

One of the basic problems which remains to be solved when a corrector system has been designed and built is how best to adjust this corrector. After the first-order optical properties have been adjusted satisfactorily, some highly sensitive technique must be employed to watch the third-order aberration disappearing as the currents or voltages are applied to the various corrector lenses. The grid shadow method, used by Deltrap and myself to measure the aberration coefficients, is not sensitive enough.

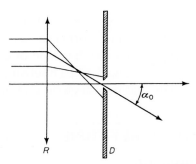

Figure 48 Meyer's arrangement to obtain a hollow beam of semi-aperture α_0. R = condenser with a large spherical aberration, D = diaphragm. Only the rays $\alpha = 0$ and $\alpha = \alpha_0$ can cross D.

Seeliger (1953) has described the method of systematic adjustment and observation which he developed. We have already mentioned this: a fine electron beam, inclined to the optic axis at a fixed angle, α_0, is progressively turned round this axis so that it traces out the aberration figure of the corrector system on the fluorescent screen, point by point; adequate magnification is provided by the projective lenses (see Fig. 20(B)). Meyer (1961) has returned to this method, but he uses a hollow conical beam with its apex in the object plane of the objective; this beam is obtained by placing a very small diaphragm in this plane, and illuminating it with an electrostatic condenser lens with very large spherical aberration. Fig. 48, which shows the paths followed by the rays which emerge from the condenser, demonstrates clearly that only paraxial rays, and those rays which are inclined to the axis at angle $\alpha_0 (\alpha_0 \sim 2 \times 10^{-2}$ rad) pass through the diaphragm. The axial beam can be used to adjust the first order optical properties, since it produces an image of the opening in the diaphragm on the final screen. This opening is obtained by placing a collodion foil on the specimen holder and strongly heating it for a very short time; holes of about 200 Å diameter can easily be obtained. The hollow beam produces a highly magnified image of the aberration figure on the screen. From the shape of this aberration figure, we can immediately deduce which of the various aperture aberrations and parasitic aberrations are present, and their relative magnitudes. By improving the centring, and methodically adjusting the octopoles, this aberration figure can be reduced to a smaller and smaller circle. This procedure is only possible up to a certain point, however, beyond which the details of the figure become interlaced and its structure vanishes. Meyer suggests that the objective should then be slightly over-focused; the Fresnel fringes which appear at the edge of the image of an aperture are then observed, and by this means, the second order astigmatism in particular can be detected (and perhaps corrected, by better alignment or with the aid of sextupoles).

With this method, it seems possible to detect a resolving power of about 4 Å, but it is by no means certain that we can go beyond this figure.

7. PERSPECTIVES FOR THE FUTURE

7.1 The Present Situation

In the course of this survey, we have given an account of the results obtained by the many research workers who have made attempts to reduce or cancel completely the aperture aberration of electrostatic or magnetic lenses.

Where the improvement of round lenses is concerned, Ruska and his colleagues have made considerable progress by choosing the excitation and geometry judiciously and then showing that a convergent magnetic objective could be used with the object immersed well inside the field. The focal length is then of the order of a millimeter, and the aberration constant is reduced to $C_3 \simeq 0.5\text{--}0.6$ mm, which corresponds in theory to a resolution of the order of one or two Ångströms in a high voltage microscope. Such lenses could also be used in X-ray microscopes, if the target were non-magnetic, but they are not suitable for devices forming an external probe; in this type of instrument, it is difficult to improve upon $C_3 \simeq 10\text{--}20$ mm, always with magnetic lenses.

The corrector systems which have been studied extensively are all based upon departure from rotational symmetry, and it does not at present seem that the other possible methods offer any serious hope of gaining any improvement. Recent experimental work has shown that C_3 can be reduced to zero, and that the additional aberrations due to mechanical imperfections can be kept below a certain level which is compatible with a valuable degree of improvement; despite this, no corrected system has yet been built which is optically of higher quality than a round lorn alone. Nevertheless, it does seem possible that with a carefully machined and well-aligned system of corrector lenses, the resolving power of a normal microscope objective can be improved by a factor of at least two or three, or of ten or twenty with a probe-forming lens. This corrector system may consist of

either   a number of weakly convergent quadrupoles and octopoles used in conjunction with a round lens

or   a highly convergent combination of quadrupoles and octopoles alone, with external foci;

the aperture aberrations to be corrected must clearly not exceed those of a round lens.

7.2 New Means of Improvement

All the methods described in this article are based upon geometrical optics, in which the aberrations of any order can be calculated once the Gaussian trajectories are known.

Very recently, however, a completely new way of correcting the third order aberration has been conceived, which stems from the work of Hoppe (1963).

The principle of the corrector which Hoppe proposes depends wholly upon the wave nature of the particles; the resolving power of an objective lens is now determined in terms of the electron density distribution in the diffraction patch which surrounds two very close object points, making due allowance for the broadening due to aperture aberration. At the exit of the lens, the wave surfaces are no longer spherical, but are given (Glaser, 1952) by

$$\rho_\theta = C_1 \sin\theta - 2C \tan\theta$$
$$\xi_\theta = -C_1 \cos\theta - C \tan^2\theta + 2C. \tag{53}$$

ρ_θ and ξ_θ are defined in Fig. 49 and

$$C = M^4 C_3.$$

As a consequence of spherical aberration, the rays leaving the perturbed surface S_p no longer arrive in phase at the point B; instead, they arrive with path differences ΔS, and this leads to a broadening of the image spot.

Suppose now that a diaphragm, consisting of concentric zones which are alternately opaque and transparent to electrons, is placed across the surface S_p; it will be possible to intercept the waves which are out of phase, just as one does with a Fresnel zone plate in classical optics, and hence to suppress the influence of the spherical aberration. For a high quality lens, the distance between the unperturbed surface S_0 and the perturbed surface S_p will be so small that the geometrical dimensions of successive rings on the diaphragm will be of the order of 10^{-2} mm, even using electron wavelengths of a fraction of an Ångström.

Hoppe (1963) has calculated the properties of various diaphragms, with nine and seventeen opaque zones (Fig. 50), and the final resolving power

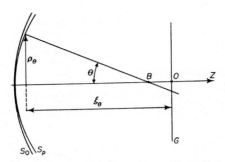

Figure 49 Perturbation of the wave surfaces by spherical aberration (round lens). $S_0 =$ unperturbed spherical surface, at the exit of a converging lens; $S_p =$ perturbed surface.

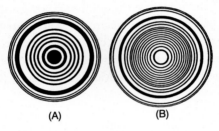

Figure 50 Zoned diaphragms calculated by Hoppe. (A) = 9 opaque zones, $d_{th} = 1.4$ Å; (B) = 14 opaque zones, $d_{th} = 1.2$ Å.

arising from the diffraction aberration: in the first case, $d_{th} = 1.4$ Å, in the second case, $d_{th} = 1.2$ Å. The diameters of the outermost rings will be 86 and 100 μ, respectively. Lenz (1963) has extended this work by calculating the electron density distributions in the vicinity of the image point; lie has also calculated the properties of a similar type of diaphragm, designed to improve the contrast between neighboring points (phase contrast).

It now seems possible to produce these diaphragms, using the recently developed techniques of micro-machining with an electron beam (Grote, Möllenstedt, & Speidel, 1965). It remains to be shown that new aberrations are not introduced by the contamination and possible deformation of the rings which electron bombardment may create. The shape of S_p will also have to be recalculated, taking the fifth order of aberration into account.

We conclude, therefore, that hope of improving electron lenses has by no means been abandoned at the present time, and that the struggle, whose beginnings coincided with those of electron optics itself, will continue dining the coming years.

REFERENCES

Archard, G. D. (1954a). *British Journal of Applied Physics*, *5*, 294.

Archard, G. D. (1954b). In *Proceedings of the international conference on electron microscopy, London* 97. London: Royal Microscopical Society.

Archard, G. D. (1955). *Proceedings of the Physical Society of London*, *68B*, 156.

Archard, G. D. (1958). *Proceedings of the Physical Society of London*, *72*, 135.

Archard, G. D., Mulvey, T., & Petrie, D. P. R. (1960). In *European conference on electron microscopy, Delft*. Delft: Netherlands Society for Electron Microscopy (p. 51).

Ash, E. A., & Gabor, D. (1955). *Proceedings of the Royal Society of London*, *228A*, 477.

Bernard, M. Y. (1952). *Comptes Rendus de l'Académie des Sciences*, *235*, 1115.

Bernard, M. Y. (1953). *Journal de Physique et le Radium*, *14*, 381, 451.

Bertein, F. (1947). *Annales de Radioélectricité*, *2*, 379.

Bertein, F. (1948). *Annales de Radioélectricité*, *3*, 49.

Brüche, E., & Scherzer, O. (1934). *Geometrische Elektronenoptik*. Berlin: Springer.

Burfoot, J. C. (1953). *Proceedings of the Physical Society of London*, *66B*, 775.

Burfoot, J. C. (1954). *Proceedings of the Physical Society of London*, *67B*, 523.

Crewe, A. V. (1965). *Conference on non-conventional electron microscopy*. Cambridge (not published).

Deltrap, J. H. (1964a). Thesis, Cambridge.

Deltrap, J. H. (1964b). In *Proceedings of the European conference on electron microscopy, Prague*. Prague: Czechoslovak Academy of Sciences (p. 45).

Deltrap, J. H., & Cosslett, V. E. (1962). In *Proceedings of the international conference on electron microscopy, Philadelphia, KK 8*. New York: Academic Press.

Dhuicq, D. (1961). Thèse 3ème Cycle. Paris.

Dhuicq, D., & Septier, A. (1959). *Comptes Rendus de l'Académie des Sciences Paris*, *249*, 2031.

Dosse, J. (1941). *Zeitschrift für Physik*, *117*, 316, 722.

Dugas, J., Durandeau, P., & Fert, C. (1961). *Revue d'Optique*, *40*, 277.

Dungey, J. W., & Hull, C. R. (1947). *Proceedings of the Physical Society of London*, *59B*, 828.

Dupouy, G., & Trinquier, J. (1962). Journées Franç. Mic. Elect. Toulouse (not published).

Dupouy, G., Perrier, F., & Trinquier, J. (1963). *Comptes Rendus de l'Académie des Sciences Paris*, *257*, 1099.

Dupouy, G., Perrier, F., & Trinquier, J. (1964). *Journal de Microscopie*, *3*, 115.

Durandeau, P., & Fert, C. (1957). *Revue d'Optique*, *36*, 205.

Dymnikov, A. D., & Yavor, S. Ya. (1964a). *Soviet Physics Technical Physics*, *8*, 639.

Dymnikov, A. D., & Yavor, S. Ya. (1964b). In *Proceedings of the European conference on electron microscopy, Prague*. Prague: Czechoslovak Academy of Sciences (p. 43).

Gabor, D. (1945a). *The electron microscope*. London: Hulton Press.

Gabor, D. (1945b). *Proceedings of the Royal Society*, *183A*, 436.

Gabor, D. (1946). *Nature (London)*, *158*, 198.

Gianola, V. F. (1950a). *Proceedings of the Physical Society of London*, *63B*, 703.

Gianola, V. F. (1950b). *Proceedings of the Physical Society of London*, *63B*, 1037.

Glaser, W. (1938). *Zeitschrift für Physik*, *109*, 700.

Glaser, W. (1940). *Zeitschrift für Physik*, *116*, 19.

Glaser, W. (1941). *Zeitschrift für Physik*, *117*, 285.

Glaser, W. (1952). *Grundlagen der Elektronenoptik*. Vienna: Springer.

Glaser, W. (1955). Farrand Optical Company, Internal Reports.

Glaser, W. (1956). *Handbuch der Physik*, *33*, 123. Berlin: Springer.

Glaser, W., & Schiske, P. (1954). *Optik*, *11*, 455.

Gobrecht, R. (1956). *Explle. Tech. Phys.*, *4*, 215.

Grivet, P., & Septier, A. (1960). *Nuclear Instruments and Methods*, *6*, 126, 243.

Grote, K. H. v., Möllenstedt, G., & Speidel, R. (1965). *Optik*, *22*, 252.

Gundert, E. (1939). *Zeitschrift für Physik*, *112*, 689.

Hanszen, K.-J. (1958a). *Zeitschrift für Naturforschung*, *13a*, 409.

Hanszen, K.-J. (1958b). *Optik*, *15*, 304.

Hast, N. (1948). *Nature (London)*, *162*, 892.

Haufe, G. (1958). *Optik*, *15*, 521.

Hawkes, P. W. (1962). In *Proceedings of the international conference on electron microscopy, Philadelphia, KK 7*. New York: Academic Press.

Hawkes, P. W. (1963). PhD Thesis. Cambridge.

Hawkes, P. W. (1965a). *Philosophical Transactions of the Royal Society*, *257*, 479.

Hawkes, P. W. (1965b). *Optik*, *22*, 349.

Heise, F. (1949). *Optik*, *5*, 479.

Hoppe, W. (1963). *Optik*, *20*, 599.

Hubert, P. (1949). *Comptes Rendus de l'Académie des Sciences Paris*, *228*, 233.

Hubert, P. (1951). *Comptes Rendus de l'Académie des Sciences Paris*, *233*, 943.

Kompfner, R. (1941). *Philosophical Magazine*, *32*, 410.

Lapostolle, P., see Regenstreif (1959). CERN 59-26, 66.

Lenz, F. (1950). *Zeitschrift für Angewandte Physik*, *2*, 448.

Lenz, F. (1951). *Annalen der Physik*, *9*, 245.

Lenz, F. (1963). *Zeitschrift für Physik*, *172*, 498.

Liebmann, G. (1949). *Proceedings of the Physical Society of London*, *62B*, 213.

Liebmann, G. (1951). *Proceedings of the Physical Society of London*, *64B*, 972.

Liebmann, G., & Grad, E. M. (1951). *Proceedings of the Physical Society of London*, *64B*, 956.

Lippert, W., & Pohlit, W. (1953). *Optik*, *10*, 447.

Mahl, H., & Recknagel, A. (1944). *Zeitschrift für Physik*, *122*, 660.

Markovich, M. G., & Tsukkerman, I. I. (1961). *Soviet Physics Technical Physics*, *5*, 1292.

Marton, L. (1939). *Physical Review*, *55*, 672.

Marton, L., & Bol, K. (1947). *Journal of Applied Physics*, *18*, 522.

Meyer, W. E. (1956). *Optik*, *13*, 86.

Meyer, W. E. (1958). *Optik*, *15*, 398.

Meyer, W. E. (1961). *Optik*, *18*, 69.

Mladjenović, M. (1953). *Bulletin Scientifique - Conseil des Académies de la RPF Yougoslavie*, *1*, 12.

Möllenstedt, G. (1956). *Optik*, *13*, 209.

Nadeau, G. (1951). *Rendiconti Accademia Nazionale dei Lincei*, *10*, 225.

Plass, G. N. (1942). *Journal of Applied Physics*, *13*, 49.

Ramberg, E. G. (1942). *Journal of Applied Physics*, *13*, 582.

Rang, O. (1949). *Optik*, *5*, 518.

Rebsch, R. (1938). *Annalen der Physik*, *31*, 551.

Rebsch, R., & Schneider, W. (1937). *Zeitschrift für Physik*, *107*, 138.

Recknagel, A. (1940). *Zeitschrift für Physik*, *117*, 67.

Reisman, E. (1957). Thesis. Cornell University.

Riecke, W. D. (1962a). In *Proceedings of the fifth international congress on electron microscopy, Philadelphia, KK 5*. New York: Academic Press.

Riecke, W. D. (1962b). *Optik*, *19*, 169.

Rüdenberg, R. (1948). *Journal of the Franklin Institute, 246*, 311, 377.

Ruska, E. (1962). In *Proceedings of the fifth international congress on electron microscopy, Philadelphia, A 1.* New York: Academic Press.

Ruska, E. (1964). *J. Micr., 3*, 357.

Ruska, E. (1965). *Optik, 22*, 319.

Scherzer, O. (1936a). *Zeitschrift für Physik, 101*, 593.

Scherzer, O. (1936b). *Zeitschrift für Physik, 101*, 23.

Scherzer, O. (1947). *Optik, 2*, 114.

Seeliger, R. (1948). *Optik, 4*, 258.

Seeliger, R. (1949). *Optik, 5*, 490.

Seeliger, R. (1951). *Optik, 8*, 311.

Seeliger, R. (1953). *Optik, 10*, 29.

Septier, A. (1957). *Comptes Rendus de l'Académie des Sciences Paris, 245*, 1406.

Septier, A. (1958). *Comptes Rendus de l'Académie des Sciences Paris, 246*, 1983.

Septier, A. (1960). CERN 60-39.

Septier, A. (1961). *Advances in Electronics and Electron Physics, 14*, 85.

Septier, A. (1962). In *Proceedings of the international conference on electron microscopy, Philadelphia, KK 10.* New York: Academic Press.

Septier, A. (1963). In *Conference on electron microscopy.* Cambridge (not published).

Septier, A., & Dhuicq, D. (1964). *Journées Franç. Mic. Electr. Strasbourg* (not published).

Septier, A., & Ruytoor, M. (1959). *Comptes Rendus de l'Académie des Sciences Paris, 249*, 2175, 2476.

Septier, A., & Van Acker, J. (1961). *Nuclear Instruments and Methods, 13*, 335.

Sturrock, P. A. (1951). *Proceedings of the Royal Society A, 210*, 269.

Sturrock, P. A. (1955). *Static and dynamic electron optics.* Cambridge University Press (p. 24).

Tretner, W. (1959). *Optik, 16*, 155.

Whitmer, R. F. (1956). *Journal of Applied Physics, 27*, 808.

Zworykin, V. K., Morton, G. A., Ramberg, E. G., Hillier, J., & Vance, A. W. (1945). *Electron optics and the electron microscope.* New York: Wiley.

INDEX

Printed in the United States
By Bookmasters